An Introduction to Calculus

With Hyperbolic Functions, Limits, Derivatives, and More

Duc Van Khanh Tran

Table of Contents

About the Author 1

Preface 2

Chapter 1: Hyperbolic Functions 4
- 1.1 Introduction to Hyperbolic Functions 5
- 1.2 Identities of Hyperbolic Functions 7
- 1.3 Inverse Hyperbolic Functions 13
- Exercise Problems 17

Chapter 2: Introduction to Limits 18
- 2.1 Definition of Limits 19
- 2.2 Existence of a Limit 20
- 2.3 Continuity and Removable Discontinuity of a Function 21
- 2.4 Limits of Rational Functions 23
- 2.5 Properties of Limits 25
- 2.6 Squeeze Theorem 27
- Exercise Problems 31

Chapter 3: Introduction to Ordinary Derivatives 32
- 3.1 Introduction 33
- 3.2 Definition of Derivatives 33
- 3.3 Basic Properties of Derivatives 38
- 3.4 Power Rule 38
- 3.5 Derivatives of Exponential Functions 39
- 3.6 Derivatives of Logarithmic Functions 40
- 3.7 Derivatives of Trigonometric Functions 41
- 3.8 Derivatives of Inverse Trigonometric Functions 42
- 3.9 Derivatives of Hyperbolic Functions 43
- 3.10 Derivatives of Inverse Hyperbolic Functions 45

3.11 Derivative of the Absolute Value Function 46
3.12 Differentiability 49
Exercise Problems 51

Chapter 4: Product Rule, Quotient Rule, and Chain Rule **52**
4.1 Introduction 53
4.2 Product Rule 53
4.3 Quotient Rule 56
4.4 Chain Rule 62
Exercise Problems 66

Chapter 5: Second Derivatives, Implicit Differentiation, and Logarithmic Differentiation **67**
5.1 Second Derivatives 68
5.2 Implicit Differentiation 69
5.3 Logarithmic Differentiation 74
Exercise Problems 76

Chapter 6: Analyzing Single Variable Functions **77**
6.1 Introduction 78
6.2 Increasing and Decreasing at a Point 78
6.3 Critical Points 80
6.4 Maxima and Minima of Functions 81
6.5 Concave Up, Concave Down, and Inflection Points 87
Exercise Problems 91

Chapter 7: Indeterminate Forms of Limits and L'Hôpital's Rule **92**
7.1 Indeterminate Forms and L'Hôpital's Rule 93
Exercise Problems 102

Chapter 8: Stirling's Formula **103**
8.1 Introduction to Factorial 104
8.2 Stirling's Formula 105
Exercise Problems 108

Chapter 9: Floor Function, Ceiling Function, and Fractional Part Function **109**
9.1 Floor Function 110
9.2 Ceiling Function 113
9.3 Fractional Part Function 115
Exercise Problems 117

Chapter 10: Some Basic Higher Derivatives **119**
10.1 Higher Than Second Derivative 120
10.2 Some General Patterns for Higher Derivatives 121
Exercise Problems 126

Chapter 11: Introduction to Summation, Product, Arithmetic Sequence, and Geometric Sequence **127**
11.1 Introduction to Summation 128
11.2 Introduction to Product 130
11.3 Arithmetic Sequence 132
11.4 Geometric Sequence 134
Exercise Problems 139

Chapter 12: Taylor Series and Maclaurin Series **140**
12.1 Taylor Series, Linear Approximation, and Quadratic Approximation 141
12.2 Maclaurin Series 142
12.3 An Interesting Sum 145
Exercise Problems 148

Chapter 13: Euler's Formula with Trigonometric Functions and Hyperbolic Functions **149**
13.1 Euler's Formula 150
13.2 Non-High-School Definitions of Sine and Cosine 151
13.3 Relations between Hyperbolic Functions and Trigonometric Functions 152
Exercise Problems 155

Chapter 14: Introduction to Partial Derivatives **156**
14.1 Introduction 157
14.2 Formal Definitions and How to Calculate 159
14.3 Multivariable Chain Rule 162
Exercise Problems 164

Answers to Exercise Problems **165**

Sources Cited in the Book **170**

Acknowledgements **172**

Reviews **174**

About the Author

Hi! My name is Duc Van Khanh Tran, and I am Vietnamese. I am currently a high school student at Brentwood Christian School in Austin, Texas, USA. I "fell in love" with mathematics when I was in 9th grade, and I have had a passion for mathematics since then.

When I was in elementary school, I studied at Morinosato Elementary School in Kanazawa, Ishikawa province, Japan for about two years. For middle school education, I studied at Le Quy Don Middle School in Ho Chi Minh City, Vietnam.

When I was in 9th grade, I came to the USA. When I first studied in a high school math class in the USA, the first thing I thought was, "This is what I learned in middle school in Vietnam!" Because I studied ahead, I felt more relaxed about the math class. Not only the math class, other classes were also more relaxing and less stressful.

Thanks to the relaxing study environment that the school offers, I began to understand and enjoy the things I studied at school. After a while, I realized that I enjoyed math the most out of all the subjects. Also, there is a math team in my current school, and the math team has made me love mathematics even more.

When I became a junior (11th grade), I came across the *@daily_math_* page on Instagram. The *@daily_math_* page is a very popular math page on Instagram, and I really liked that page and started following it. Inspired by the *@daily_math_* page, I also created a math page on Instagram called *@dvkt_math* with about 25,000 followers currently. I created the math page *@dvkt_math* because I wanted to share mathematics knowledge with many people, and that is also the reason why I am writing this book.

Preface

My inspiration for writing this book is nothing impressive, but I will talk about it anyway. The admin of the *@daily_math_* page on Instagram, Hamza Alsamraee, is the author of an advanced calculus book called *Advanced Calculus Explored*, and he is only one year older than I am. When I talked to my dad about a guy who is only one year older writing an advanced calculus book, my dad told me, "Why don't you write a book, too?" So, I just went along with the flow, and here I am writing an introductory calculus book.

Have you ever heard of limits? What about derivatives? Have you ever seen this before?

$$f'(x) = \lim_{h \to 0} \frac{f(x+h) - f(x)}{h}.$$

If you have never heard of these terms or seen the formula above, it is completely fine. This book is intended for those who have never learned calculus before. If you already knew these terms and formula or have learned a little bit of calculus before, do not leave yet. This book starts with some very basic topics, but then leads to some relatively more advanced topics by the end of the book - with even a brief introduction to partial derivatives in the last chapter.

With that said, this book introduces calculus in a little more challenging way than normal. This book does not stick with the usual curriculum of AP Calculus in high school or Calculus 1 or 2 in college in the US at all. Also, this book focuses on differential calculus and does not cover topics related to integrals. In most parts of this book, we will stay in the realm of real numbers, except for chapter 13 where we will make a short journey into the realm of complex numbers.

The prerequisite for this book is pre-calculus in the USA high school education or its equivalent. I assume that the readers know how to do simple algebraic operations (addition, subtraction, multiplication, division, exponents, and logarithms), conic sections (hyperbola, circle, parabola), trigonometry, and some basic concepts related to functions including how to graph some basic functions (exponential functions, logarithmic functions, rational functions, etc.). A little bit of basic geometry is used in chapter 2 as well, and some very basic knowledge about complex numbers is necessary for chapter 13.

Also, you might be familiar with using the notation log for natural logarithm, but in this book the notation ln is used for natural logarithm. As for the notations used in this book in general, there might be many notations you are not quite familiar with yet, but do not worry - I will explain those unfamiliar notations when they are used in this book.

As for the graphs in this book, they are generated using the online graphing calculator *desmos.com* (except for the graph of Weierstrass function in section 3.12). Desmos is not a perfect graphing calculator, and it has some deficiencies. When there is some deficiency that affects important features of the graphs, I will explain those features below the graphs. The graph of Weierstrass function in section 3.12 is taken from the Wikipedia article about Weierstrass function.

Enjoy reading!

Duc Van Khanh Tran
Texas, USA, 2020

Chapter 1

Hyperbolic Functions

1.1 Introduction to Hyperbolic Functions

In this chapter, I would like to give you a brief introduction to hyperbolic functions. Hyperbolic functions are analogs of trigonometric functions defined on the unit hyperbola , $x^2 - y^2 = 1$, instead of the unit circle, $x^2 + y^2 = 1$. Here is a list of hyperbolic functions with their domains (D) and ranges (R).

Hyperbolic Functions

$\sinh x = \dfrac{e^x - e^{-x}}{2}$	D: $(-\infty, \infty)$	R: $(-\infty, \infty)$
$\cosh x = \dfrac{e^x + e^{-x}}{2}$	D: $(-\infty, \infty)$	R: $[1, \infty)$
$\tanh x = \dfrac{\sinh x}{\cosh x}$	D: $(-\infty, \infty)$	R: $(-1, 1)$
$\coth x = \dfrac{1}{\tanh x}$	D: $(-\infty, 0) \cup (0, \infty)$	R: $(-\infty, -1) \cup (1, \infty)$
$\operatorname{sech} x = \dfrac{1}{\cosh x}$	D: $(-\infty, \infty)$	R: $(0, 1]$
$\operatorname{csch} x = \dfrac{1}{\sinh x}$	D: $(-\infty, 0) \cup (0, \infty)$	R: $(-\infty, 0) \cup (0, \infty)$

Note that hyperbolic functions are very similar to trigonometric functions. For example, for trigonometric functions, we have

$$\tan x = \frac{\sin x}{\cos x}.$$

Similarly, for hyperbolic functions, we have

$$\tanh x = \frac{\sinh x}{\cosh x}.$$

Also, for trigonometric functions, we have

$$\csc x = \frac{1}{\sin x}.$$

Similarly, we also have

$$\operatorname{csch} x = \frac{1}{\sinh x}$$

for hyperbolic functions. Here are a few examples to familiarize yourself with hyperbolic functions.

Example 1.1-1: Find the value of $\sinh 1 + \cosh 1$.
By definition,

$$\sinh x = \frac{e^x - e^{-x}}{2} \text{ and } \cosh x = \frac{e^x + e^{-x}}{2}.$$

Thus, we have

$$\sinh x + \cosh x = \frac{e^x - e^{-x}}{2} + \frac{e^x + e^{-x}}{2} = e^x.$$

Plugging in $x = 1$,

$$\sinh 1 + \cosh 1 = e^1 = e.$$

Example 1.1-2: Find the value of $\operatorname{sech} 2 + \operatorname{csch} 2$.
By definition,

$$\operatorname{sech} x = \frac{1}{\cosh x} = \frac{2}{e^x + e^{-x}},$$

and

$$\operatorname{csch} x = \frac{1}{\sinh x} = \frac{2}{e^x - e^{-x}}.$$

Thus, we have

$$\operatorname{sech} x + \operatorname{csch} x = \frac{2}{e^x + e^{-x}} + \frac{2}{e^x - e^{-x}} = \frac{4e^x}{e^{2x} - e^{-2x}}.$$

Plugging in $x = 2$,

$$\operatorname{sech} 2 + \operatorname{csch} 2 = \frac{4e^2}{e^4 - e^{-4}} = \frac{4e^6}{e^8 - 1}.$$

Example 1.1-3: Find the value of $\tanh 3 + \coth 3$.
By definition,

$$\tanh x = \frac{\sinh x}{\cosh x} = \frac{e^x - e^{-x}}{e^x + e^{-x}}$$

and

$$\coth x = \frac{1}{\tanh x} = \frac{e^x + e^{-x}}{e^x - e^{-x}}.$$

Thus, we have

$$\tanh x + \coth x = \frac{e^x - e^{-x}}{e^x + e^{-x}} + \frac{e^x + e^{-x}}{e^x - e^{-x}} = \frac{2e^{2x} + 2e^{-2x}}{e^{2x} - e^{-2x}}.$$

Plugging in $x = 3$,

$$\tanh 3 + \coth 3 = \frac{2e^6 + 2e^{-6}}{e^6 - e^{-6}} = \frac{2e^{12} + 2}{e^{12} - 1}.$$

1.2 Identities of Hyperbolic Functions

After getting to know hyperbolic functions, let us learn the identities of hyperbolic functions. You will be surprised by how similar they are to the identities of trigonometric functions.

First, let us see which functions are odd and which functions are even.

Even Hyperbolic Functions

$$\cosh(-x) = \cosh x$$
$$\operatorname{sech}(-x) = \operatorname{sech} x$$

Odd Hyperbolic Functions

$$\sinh(-x) = -\sinh x$$
$$\tanh(-x) = -\tanh x$$
$$\coth(-x) = -\coth x$$
$$\operatorname{csch}(-x) = -\operatorname{csch} x$$

Note that the odd and even behaviors of hyperbolic functions are very similar to those of trigonometric functions. For example, $\cos x$ is an even function, and $\cosh x$ is also an even function. In addition to that, $\sin x$ is an odd function, and $\sinh x$ is also an odd function.

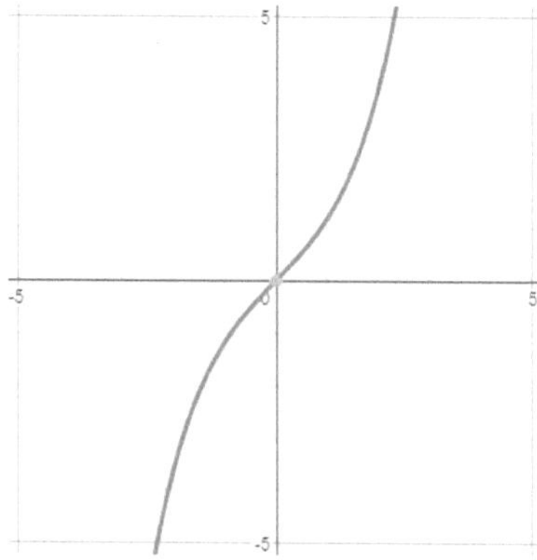

Figure 1.1: Graph of $f(x) = \sinh x$

Now let us talk about the even-odd decomposition of a function. Any function $f(x)$ can be written as a sum of an even function $E(x)$ and an odd function $O(x)$:

$$f(x) = E(x) + O(x).$$

Substituting $x \to -x$,

$$f(-x) = E(-x) + O(-x).$$

Since $E(-x) = E(x)$ and $O(-x) = -O(x)$,

$$f(-x) = E(x) - O(x).$$

Then, we can see that

$$E(x) = \frac{f(x) + f(-x)}{2}$$

and

$$O(x) = \frac{f(x) - f(-x)}{2}.$$

So, what is the even-odd decomposition of $f(x) = e^x$? Using the formulae above,

$$E(x) = \frac{e^x + e^{-x}}{2} = \cosh x,$$

and

$$O(x) = \frac{e^x - e^{-x}}{2} = \sinh x.$$

So, $\cosh x$ is the even part of e^x, and $\sinh x$ is the odd part. We also proved in example 1.1-1 that $\cosh x + \sinh x = e^x$.

Next, let us take a look at the identities of squares of hyperbolic functions.

Hyperbolic Function Identities

$$\cosh^2 x - \sinh^2 x = 1$$
$$\tanh^2 x + \operatorname{sech}^2 x = 1$$
$$\coth^2 x - \operatorname{csch}^2 x = 1$$

Here, we need to be careful with the sign. For trigonometric functions, we have the identity

$$\cos^2 x + \sin^2 x = 1.$$

However, for hyperbolic functions, we have

$$\cosh^2 x - \sinh^2 x = 1 \tag{1.1}$$

with a negative sign in between instead of a positive sign. Also, from (1.1), we can derive the other two identities.

Dividing both sides of (1.1) by $\cosh^2 x$,

$$\begin{aligned}\frac{\cosh^2 x}{\cosh^2 x} - \frac{\sinh^2 x}{\cosh^2 x} &= \frac{1}{\cosh^2 x}\\ 1 - \tanh^2 x &= \operatorname{sech}^2 x.\end{aligned}$$

Adding $\tanh^2 x$ to both sides, we get

$$\tanh^2 x + \operatorname{sech}^2 x = 1.$$

Now dividing both sides of (1.1) by $\sinh^2 x$,

$$\begin{aligned}\frac{\cosh^2 x}{\sinh^2 x} - \frac{\sinh^2 x}{\sinh^2 x} &= \frac{1}{\sinh^2 x}\\ \coth^2 x - 1 &= \operatorname{csch}^2 x.\end{aligned}$$

Adding $1 - \operatorname{csch}^2 x$ to both sides, we get

$$\coth^2 x - \operatorname{csch}^2 x = 1.$$

Next, let us learn the sum and difference identities for hyperbolic functions.

Hyperbolic Function Identities

$$\begin{aligned}\sinh(x \pm y) &= \sinh x \cosh y \pm \cosh x \sinh y\\ \cosh(x \pm y) &= \cosh x \cosh y \pm \sinh x \sinh y\\ \tanh(x \pm y) &= \frac{\tanh x \pm \tanh y}{1 \pm \tanh x \tanh y}\end{aligned}$$

Note that the formula for $\sinh(x \pm y)$ is very similar to the formula for $\sin(x \pm y)$ since

$$\sin(x \pm y) = \sin x \cos y \pm \cos x \sin y.$$

However, we need to be careful about the signs for the formulae of $\cosh(x \pm y)$ and $\tanh(x \pm y)$. Since we have

$$\cos(x \pm y) = \cos x \cos y \mp \sin x \sin y$$

and

$$\tan(x \pm y) = \frac{\tan x \pm \tan y}{1 \mp \tan x \tan y},$$

there is a difference in the signs. Here are some examples using some of the identities we have learned so far.

Example 1.2-1: Given that $\sinh x = \sqrt{2}$ and $\sinh y = 3$, find the value of $\cosh(x+y)$.
To solve this problem, we will use the formula

$$\cosh(x+y) = \cosh x \cosh y + \sinh x \sinh y. \tag{1.2}$$

The values of $\sinh x$ and $\sinh y$ are given, and we need to find the values of $\cosh x$ and $\cosh y$. Using the identity

$$\cosh^2 x - \sinh^2 x = 1,$$

we obtain

$$\begin{aligned}\cosh^2 x - \left(\sqrt{2}\right)^2 &= 1\\ \cosh^2 x &= 3.\end{aligned}$$

Note that $\cosh x \geq 1$ for all real x.

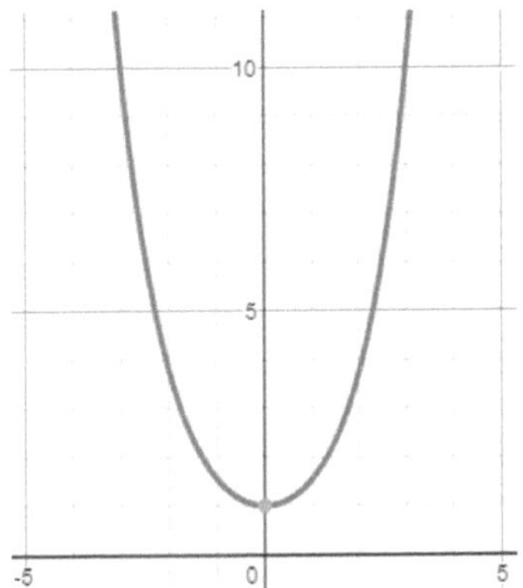

Figure 1.2: Graph of $f(x) = \cosh x$

Thus, we have

$$\cosh x = \sqrt{3}.$$

Similarly,

$$\begin{aligned}\cosh^2 y - \sinh^2 y &= 1\\ \cosh^2 y - 3^2 &= 1\\ \cosh^2 y &= 10\\ \cosh y &= \sqrt{10}.\end{aligned}$$

Finally, plugging the values of $\cosh x$, $\cosh y$, $\sinh x$, and $\sinh y$ into (1.2),

$$\cosh(x+y) = \sqrt{3} \times \sqrt{10} + \sqrt{2} \times 3 = \sqrt{30} + 3\sqrt{2}.$$

Example 1.2-2: Given that $\operatorname{sech} x = \frac{1}{2}$ and $\operatorname{sech} y = \frac{1}{4}$, if $x > 0$ and $y > 0$, what is the value of $\tanh(x-y)$?
To solve this problem, we will use the formula

$$\tanh(x-y) = \frac{\tanh x - \tanh y}{1 - \tanh x \tanh y}. \tag{1.3}$$

We need to find the values of $\tanh x$ and $\tanh y$ using the given values for $\operatorname{sech} x$ $\operatorname{sech} y$. Using the identity

$$\tanh^2 x + \operatorname{sech}^2 x = 1,$$

we obtain

$$\tanh^2 x + \left(\frac{1}{2}\right)^2 = 1$$
$$\tanh^2 x = \frac{3}{4}.$$

Note that $\tanh x > 0$ if $x > 0$.

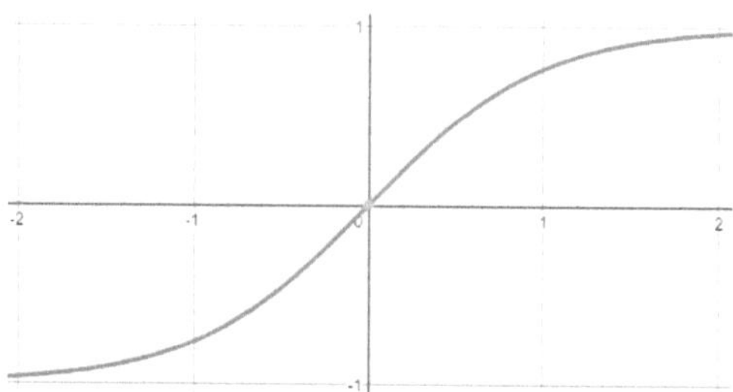

Figure 1.3: Graph of $f(x) = \tanh x$

Thus, we have

$$\tanh x = \frac{\sqrt{3}}{2}.$$

Similarly,

$$\tanh^2 y + \left(\frac{1}{4}\right)^2 = 1$$
$$\tanh^2 y = \frac{15}{16}$$
$$\tanh y = \frac{\sqrt{15}}{4}.$$

Finally, plugging the values of $\tanh x$ and $\tanh y$ into (1.3),

$$\tanh(x-y) = \frac{\frac{\sqrt{3}}{2} - \frac{\sqrt{15}}{4}}{1 - \frac{\sqrt{3}}{2} \times \frac{\sqrt{15}}{4}} = \frac{2\sqrt{3} - 4\sqrt{15}}{19}.$$

Next, let us take a look at the formulae for $\sinh(2x)$ and $\cosh(2x)$.

Hyperbolic Function Identities

$$\sinh(2x) = 2\sinh x \cosh x$$
$$\cosh(2x) = \cosh^2 x + \sinh^2 x$$

Note that the formula for $\sinh(2x)$ is very similar to the formula for $\sin(2x)$ since

$$\sin(2x) = 2\sin x \cos x.$$

For the formula of $\cosh(2x)$, we need to be careful about the signs because the formula of $\cos(2x)$ is

$$\cos(2x) = \cos^2 x - \sin^2 x$$

with a negative sign in between instead of a positive sign.

Example 1.2-3: Given that $\cosh x = 2$, find the value of $\cosh(2x) + 1$. To solve this problem, we will use the identities

$$\cosh(2x) = \cosh^2 x + \sinh^2 x \tag{1.4}$$

and

$$\cosh^2 x - \sinh^2 x = 1. \tag{1.5}$$

Adding (1.4) and (1.5) together, we get

$$\cosh(2x) + 1 = \cosh^2 x + \sinh^2 x + \cosh^2 x - \sinh^2 x = 2\cosh^2 x.$$

Since we are given that $\cosh x = 2$,

$$\cosh(2x) + 1 = 2 \times 2^2 = 8.$$

Finally, let us learn about the power reduction formulae of hyperbolic functions.

Power Reduction Formulae

$$\sinh^2 x = \frac{\cosh(2x) - 1}{2}$$
$$\cosh^2 x = \frac{\cosh(2x) + 1}{2}$$

Note that the power reduction formula for $\cosh^2 x$ is very similar to that of $\cos^2 x$ since

$$\cos^2 x = \frac{\cos(2x) + 1}{2}.$$

However, we need to be careful about the power reduction formula for $\sinh^2 x$. Since we have

$$\sin^2 x = \frac{1 - \cos(2x)}{2},$$

there is a small difference between the signs of the power reduction formula of $\sin^2 x$ and those of the power reduction formula of $\sinh^2 x$.

1.3 Inverse Hyperbolic Functions

Now it is time to learn about the inverse hyperbolic functions. Here is a list of inverse hyperbolic functions.

Inverse Hyperbolic Functions

$$\sinh^{-1} x = \ln\left(x + \sqrt{x^2 + 1}\right)$$
$$\cosh^{-1} x = \ln\left(x + \sqrt{x^2 - 1}\right)$$
$$\tanh^{-1} x = \frac{1}{2}\left[\ln(1 + x) - \ln(1 - x)\right]$$
$$\coth^{-1} x = \tanh^{-1}\left(\frac{1}{x}\right)$$
$$\operatorname{sech}^{-1} x = \cosh^{-1}\left(\frac{1}{x}\right)$$
$$\operatorname{csch}^{-1} x = \sinh^{-1}\left(\frac{1}{x}\right)$$

For most of the hyperbolic functions, their domains become the ranges of their inverse functions, and their ranges become the domains of their inverse functions. For example, $\tanh x$ has the domain $(-\infty, \infty)$ and range $(-1, 1)$, so $\tanh^{-1} x$ has the domain $(-1, 1)$ and range $(-\infty, \infty)$. The "switching" of domain and range also applies for $\sinh^{-1} x, \coth^{-1} x$, and $\operatorname{csch}^{-1} x$.

However, it does not apply for $\cosh^{-1} x$ and $\operatorname{sech}^{-1} x$. The domain of $\cosh x$ is $(-\infty, \infty)$, but the range of $\cosh^{-1} x$ is $[0, \infty)$ instead of $(-\infty, \infty)$. Similarly, the range of $\operatorname{sech}^{-1} x$ is $[0, \infty)$ while the domain of $\operatorname{sech} x$ is $(-\infty, \infty)$. The reason why we have a change in the ranges of $\cosh^{-1} x$ and $\operatorname{sech}^{-1} x$ is because we choose the positive branch for the sake of convenience. Also, if the ranges of $\cosh^{-1} x$ and $\operatorname{sech}^{-1} x$ are $(-\infty, \infty)$, $\cosh^{-1} x$ and $\operatorname{sech}^{-1} x$ would not be functions because they would fail the vertical line test. (See the graphs of $\cosh^{-1} x$ and $\operatorname{sech}^{-1} x$ in figure 1.4 and figure 1.5.)

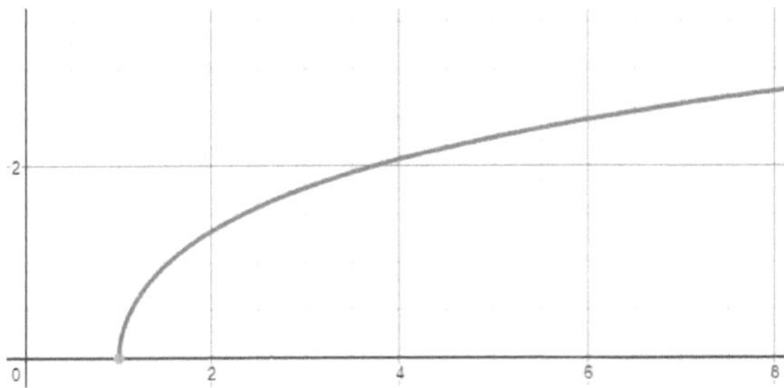

Figure 1.4: Graph of $f(x) = \cosh^{-1} x$

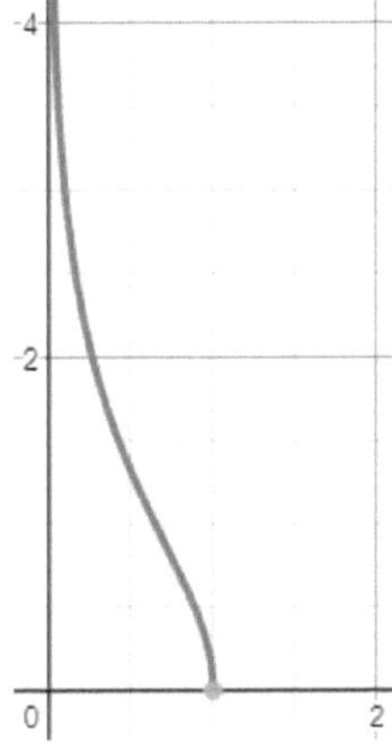

Figure 1.5: Graph of $f(x) = \operatorname{sech}^{-1} x$

Sometimes, it is useful to remember the formulae of inverse hyperbolic functions. However, we do not need to remember the formulae of inverse hyperbolic functions to find the values of inverse hyperbolic functions. Here are some examples.

Example 1.3-1: Find the value of $\sinh^{-1}(2)$.
In this problem, we are finding the value of x such that $\sinh x = 2$, or

$$\frac{e^x - e^{-x}}{2} = 2. \tag{1.6}$$

Multiplying both sides of (1.6) by 2 and then subtracting 4 from both sides,

$$\begin{aligned} e^x - e^{-x} &= 4 \\ e^x - e^{-x} - 4 &= 0. \end{aligned}$$

Multiplying both sides by e^x,

$$e^{2x} - 4e^x - 1 = 0.$$

Solving for e^x using quadratic formula gives us

$$e^x = 2 \pm \sqrt{5}.$$

Since the value of e^x must be positive,

$$e^x = 2 + \sqrt{5}.$$

Taking natural logarithm of both sides,

$$x = \ln\left(2 + \sqrt{5}\right).$$

Therefore, our final answer is

$$\sinh^{-1}(2) = \ln\left(2 + \sqrt{5}\right).$$

Example 1.3-2: Find the value of $\cosh^{-1}(3)$.
In this problem, we are finding the value of x such that $\cosh x = 3$, or

$$\frac{e^x + e^{-x}}{2} = 3. \tag{1.7}$$

Multiplying both sides of (1.7) by 2 and then subtracting 6 from both sides,

$$\begin{aligned} e^x + e^{-x} &= 6 \\ e^x + e^{-x} - 6 &= 0. \end{aligned}$$

Multiplying both sides by e^x,

$$e^{2x} - 6e^x + 1 = 0.$$

Solving for e^x using quadratic formula gives us

$$e^x = 3 \pm 2\sqrt{2}.$$

Taking natural logarithm of both sides,

$$x = \ln\left(3 \pm 2\sqrt{2}\right).$$

However, x has to be positive because the range of inverse hyperbolic cosine function is $[0, \infty)$, so $x = \ln\left(3 - 2\sqrt{2}\right)$ is an extraneous solution. Therefore, our final answer is

$$\cosh^{-1}(3) = \ln\left(3 + 2\sqrt{2}\right).$$

Example 1.3-3: Find the value of $\tanh^{-1}\left(\frac{1}{2}\right)$.
In this problem, we are finding the value of x such that $\tanh x = \frac{1}{2}$, or

$$\frac{e^x - e^{-x}}{e^x + e^{-x}} = \frac{1}{2}. \tag{1.8}$$

Multiplying both sides of (1.8) by $2\left(e^x + e^{-x}\right)$ and then subtracting $e^x + e^{-x}$ from both sides,

$$\begin{aligned} 2e^x - 2e^{-x} &= e^x + e^{-x} \\ e^x - 3e^{-x} &= 0. \end{aligned}$$

Multiplying both sides by e^x,

$$e^{2x} - 3 = 0.$$

Solving for e^x,

$$\begin{aligned} e^{2x} &= 3 \\ e^x &= \pm\sqrt{3}. \end{aligned}$$

Since the value of e^x must be positive,

$$e^x = \sqrt{3}.$$

Taking natural logarithm of both sides,

$$x = \ln\left(\sqrt{3}\right) = \frac{\ln 3}{2}.$$

Therefore, our final answer is

$$\tanh^{-1}\left(\frac{1}{2}\right) = \frac{\ln 3}{2}.$$

Exercise Problems

1. Given that $\sinh x = 2$, find:
   a) $\cosh x$,
   b) $\sinh(2x)$,
   c) $\cosh(2x)$.

2. Given that $\cosh(2x) = 2$ and $x < 0$, find $\sinh x$.

3. Find $\coth^{-1}(3)$.

4. Find $\operatorname{sech}^{-1}\left(\frac{1}{2}\right)$.

5. Find $\operatorname{csch}^{-1}\left(\frac{1}{4}\right)$.

Chapter 2

Introduction to Limits

2.1 Definition of Limits

To begin this chapter, I would like to introduce the epsilon-delta definition for a limit of a function. The Greek letter ε is epsilon, and the Greek letter δ is delta.

Epsilon-Delta Definition of a Limit
Let $f(x)$ be a function defined on an open interval around x_0. $f(x_0)$ can be undefined. We say that the limit of $f(x)$ as x approaches x_0 is L, i.e.

$$\lim_{x \to x_0} f(x) = L,$$

if for every $\varepsilon > 0$ there exists $\delta > 0$ such that for all x

$$0 < |x - x_0| < \delta \implies |f(x) - L| < \varepsilon.$$

This is the formal, rigorous definition of a limit which allows rigorous discussion of limits. However, since this book is intended for calculus beginners, we will not go over this definition in depth.

Intuitively, the limit of $f(x)$ as x approaches x_0 is the value that $f(x)$ approaches as x approaches x_0. That value does not necessarily have to be $f(x_0)$.

The expression

$$\lim_{x \to x_0} f(x)$$

is the limit of $f(x)$ as x approaches x_0 from both sides. It is a two-sided limit. Now let us take a look at the expressions for one-sided limits from the right and from the left.

One-Sided Limits
The expression for the limit of $f(x)$ as x approaches x_0 from the right is

$$\lim_{x \to x_0^+} f(x),$$

and the expression for the limit of $f(x)$ as x approaches x_0 from the left is

$$\lim_{x \to x_0^-} f(x).$$

A limit of a function can be finite, infinite, or does not exist (DNE). In the next section, we will discuss the existence of a limit.

2.2 Existence of a Limit

Here is how we can tell if a limit exists.

Existence of a Limit

If $\lim_{x \to x_0^-} f(x)$ and $\lim_{x \to x_0^+} f(x)$ exist and

$$\lim_{x \to x_0^-} f(x) = \lim_{x \to x_0^+} f(x),$$

then the limit

$$\lim_{x \to x_0} f(x)$$

exists, and

$$\lim_{x \to x_0^-} f(x) = \lim_{x \to x_0^+} f(x) = \lim_{x \to x_0} f(x).$$

Let me explain this to you through an example. For example, the limit

$$\lim_{x \to 0} \frac{1}{x} \tag{2.1}$$

does not exist. Let us take a look at the graph of $f(x) = \frac{1}{x}$.

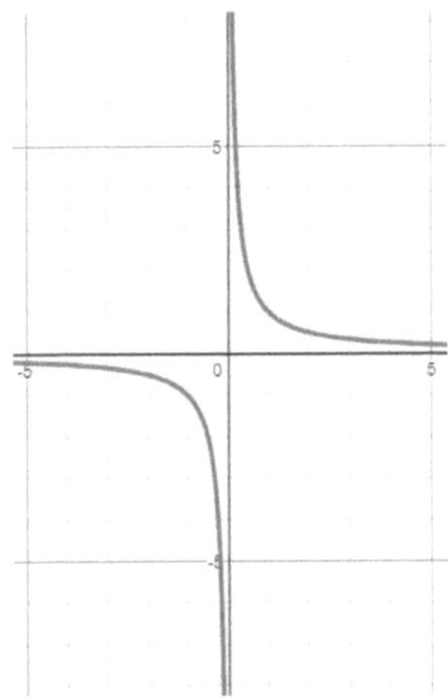

Figure 2.1: Graph of $f(x) = \frac{1}{x}$

We can see that

$$\lim_{x \to 0^+} \frac{1}{x} = \infty \text{ and } \lim_{x \to 0^-} \frac{1}{x} = -\infty.$$

Thus,

$$\lim_{x\to 0^+} \frac{1}{x} \neq \lim_{x\to 0^-} \frac{1}{x},$$

and the limit in (2.1) does not exist.

Another example we can use is

$$\lim_{x\to 0} \ln x, \tag{2.2}$$

which does not exist. Let us take a look at the graph of $f(x) = \ln x$.

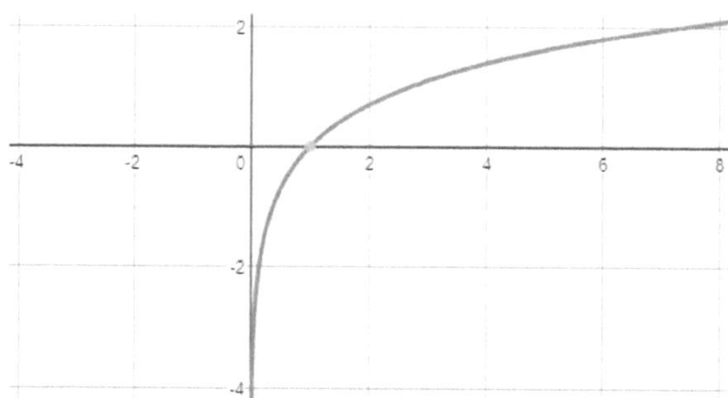

Figure 2.2: Graph of $f(x) = \ln x$

We can see that

$$\lim_{x\to 0^+} \ln x = -\infty \text{ and } \lim_{x\to 0^-} \ln x = \text{DNE}.$$

Thus,

$$\lim_{x\to 0^+} \ln x \neq \lim_{x\to 0^-} \ln x,$$

and the limit in (2.2) does not exist.

2.3 Continuity and Removable Discontinuity of a Function

First, I would like to briefly introduce the relationship between limits and continuity of a function.

> **Continuity of a Function**
> A function $f(x)$ is continuous at $x = a$ if and only if $f(a)$ is defined and
>
> $$\lim_{x\to a} f(x) = f(a).$$

Let us take a look at a basic example. We know that the function

$$f(x) = x^2$$

is continuous at $x = 1$. Then, we know that $f(1)$ is defined and that

$$\lim_{x \to 1} x^2 = 1^2 = 1.$$

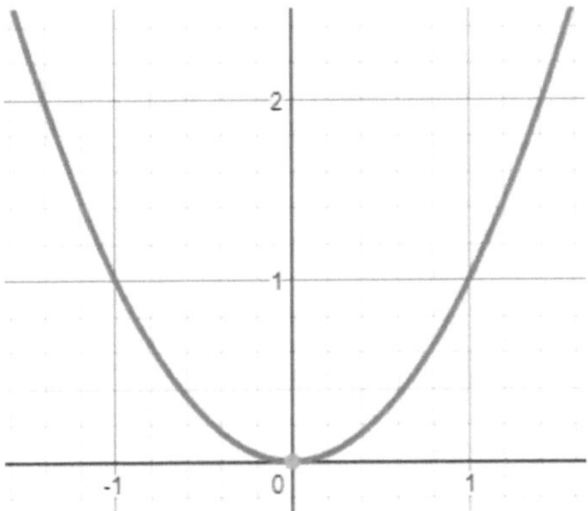

Figure 2.3: Graph of $f(x) = x^2$

Next, let us consider the following question.

Example 2.3-1: Is the function

$$f(x) = \frac{x-2}{x^2 - 7x + 10}$$

continuous at $x = 2$?
First, let us find

$$\lim_{x \to 2} f(x).$$

We have

$$\lim_{x \to 2} \frac{x-2}{x^2 - 7x + 10} = \lim_{x \to 2} \frac{x-2}{(x-2)(x-5)} = \lim_{x \to 2} \frac{1}{x-5}.$$

Plugging in $x = 2$,

$$\lim_{x \to 2} \frac{1}{x-5} = \frac{-1}{3}.$$

Now, let us see if $f(2)$ is defined. Note that the denominator of $f(x)$ is $(x-2)(x-5)$. So, if we plug in $x = 2$, the denominator will be 0, and thus $f(2)$ is undefined.
Therefore, $f(x)$ is not continuous at $x = 2$, or more specifically, $f(x)$ has removable discontinuity at the point $\left(2, \frac{-1}{3}\right)$.

Removable Discontinuity
If

$$\lim_{x\to x_0} f(x) = a$$

for finite a, and $f(x_0)$ is undefined or defined to be a different value from a, then we say that $f(x)$ has removable discontinuity, which appears as a hole in the graph, at the point (x_0, a), and vice versa.

Intuitively, a removable discontinuity of a function $f(x)$ is the point where $f(x)$ is discontinuous but can be continuous if we "fill in the hole" by defining

$$f(x_0) = a.$$

2.4 Limits of Rational Functions

In this section, we will discuss the limits of rational functions as x approaches ∞ or $-\infty$.

Limits of Rational Functions
Let $P(x)$ be a polynomial of degree n with leading coefficient a and $Q(x)$ be a polynomial of degree m with leading coefficient b.
We have

$$\lim_{x\to\infty} \frac{P(x)}{Q(x)} = \begin{cases} 0 & \text{if } n < m \\ \frac{a}{b} & \text{if } n = m \\ \infty & \text{if } n > m \end{cases}$$

and

$$\lim_{x\to-\infty} \frac{P(x)}{Q(x)} = \begin{cases} 0 & \text{if } n < m \\ \frac{a}{b} & \text{if } n = m \\ \infty & \text{if } n > m \text{ and } n - m \text{ is even} \\ -\infty & \text{if } n > m \text{ and } n - m \text{ is odd} \end{cases}$$

for the limits of rational function $\dfrac{P(x)}{Q(x)}$.

Note that when $n < m$ or $n = m$,

$$\lim_{x\to\infty} \frac{P(x)}{Q(x)} = \lim_{x\to-\infty} \frac{P(x)}{Q(x)},$$

and both limits are finite. In these cases, $\frac{P(x)}{Q(x)}$ has horizontal asymptote, which is the value that $\frac{P(x)}{Q(x)}$ approaches at infinity.

> **Horizontal Asymptote of Rational Function**
> Let $P(x)$ be a polynomial of degree n with leading coefficient a and $Q(x)$ be a polynomial of degree m with leading coefficient b.
> If $n < m$ or $n = m$, then $\frac{P(x)}{Q(x)}$ has horizontal asymptote at
>
> $$y = \lim_{x\to\infty} \frac{P(x)}{Q(x)} = \lim_{x\to-\infty} \frac{P(x)}{Q(x)}.$$
>
> If $n > m$, then $\frac{P(x)}{Q(x)}$ does not have horizontal asymptote.

Example 2.4-1: Find the horizontal asymptote of $f(x) = \frac{1}{x}$.
The limits of $f(x)$ are

$$\lim_{x\to-\infty} \frac{1}{x} = \lim_{x\to\infty} \frac{1}{x} = 0.$$

Thus, the horizontal asymptote of $f(x) = \frac{1}{x}$ is

$$y = 0.$$

Example 2.4-2: Find the horizontal asymptote of $f(x) = \frac{x^3 + 2x^2 + 1}{5x^3 + x^2 - 4x}$.
The limits of $f(x)$ are

$$\lim_{x\to-\infty} \frac{x^3 + 2x^2 + 1}{5x^3 + x^2 - 4x} = \lim_{x\to\infty} \frac{x^3 + 2x^2 + 1}{5x^3 + x^2 - 4x} = \frac{1}{5}.$$

Thus, the horizontal asymptote of $f(x) = \frac{x^3 + 2x^2 + 1}{5x^3 + x^2 - 4x}$ is

$$y = \frac{1}{5}.$$

We can extend this concept of horizontal asymptote to any funtion $f(x)$ other than the rational functions. If $f(x)$ does not go to $\pm\infty$ as x goes to $\pm\infty$, then $f(x)$ has horizontal asymptote at $y = \lim_{x\to\infty} f(x)$ and $y = \lim_{x\to-\infty} f(x)$.

2.5 Properties of Limits

Here is a list of properties of limits.

Properties of Limits
Let $f(x)$ and $g(x)$ be functions in terms of x. Assuming that $\lim_{x \to a} f(x)$ and $\lim_{x \to a} g(x)$ exist,

$$\lim_{x \to a} (f(x) \pm g(x)) = \lim_{x \to a} f(x) \pm \lim_{x \to a} g(x)$$

$$\lim_{x \to a} f(x)g(x) = \lim_{x \to a} f(x) \times \lim_{x \to a} g(x)$$

$$\lim_{x \to a} \frac{f(x)}{g(x)} = \frac{\lim_{x \to a} f(x)}{\lim_{x \to a} g(x)} \qquad \text{if } g(x) \neq 0$$

$$\lim_{x \to a} f(x)^{g(x)} = \left(\lim_{x \to a} f(x)\right)^{\lim_{x \to a} g(x)}$$

Now let us dive into a few examples to familiarize ourselves with these properties.

Example 2.5-1: Given that $\lim_{x \to 0^+} x \ln x = 0$ and $\lim_{x \to 0^+} x^x = 1$, find

$$\lim_{x \to 0^+} (x \ln x + x^x).$$

Using the property

$$\lim_{x \to a} (f(x) + g(x)) = \lim_{x \to a} f(x) + \lim_{x \to a} g(x),$$

we have

$$\lim_{x \to 0^+} (x \ln x + x^x) = \lim_{x \to 0^+} x \ln x + \lim_{x \to 0^+} x^x = 0 + 1 = 1.$$

From the graph in figure 2.4, we can see that indeed

$$\lim_{x \to 0^+} (x \ln x + x^x) = 1.$$

In the example above, the limits $\lim_{x \to 0^+} x \ln x$ and $\lim_{x \to 0^+} x^x$ are given, but we can evaluate these limits by ourselves. However, to evaluate these limits, we need to understand the basic concepts of derivatives (which will be discussed in chapters 3-4) and be familiar with indeterminate forms of limits and L'Hôpital's rule (which will be discussed in chapter 7).

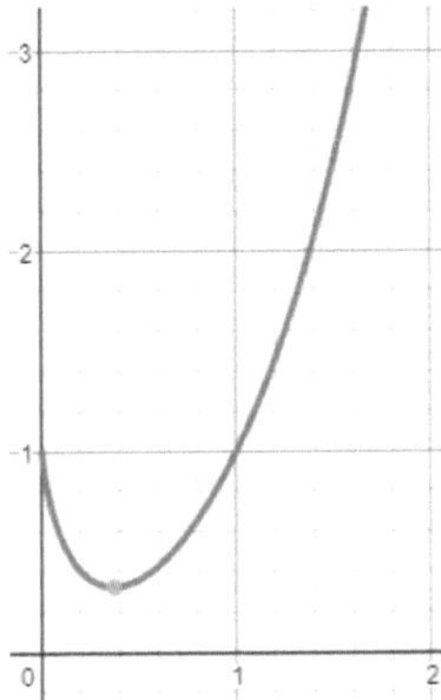

Figure 2.4: Graph of $f(x) = x \ln x + x^x$

Example 2.5-2 Find $\lim\limits_{x \to 0} 4 \cos x$.
Using the property

$$\lim_{x \to a} f(x)g(x) = \lim_{x \to a} f(x) \times \lim_{x \to a} g(x),$$

we have

$$\lim_{x \to 0} 4 \cos x = \lim_{x \to 0} 4 \times \lim_{x \to 0} \cos x.$$

Note that

$$\lim_{x \to 0} 4 = 4$$

since 4 is independent of x. Plugging in $x = 0$,

$$\lim_{x \to 0} \cos x = \cos 0 = 1.$$

Therefore, our final answer is

$$\lim_{x \to 0} 4 \cos x = \lim_{x \to 0} 4 \times \lim_{x \to 0} \cos x = 4 \times 1 = 4.$$

Example 2.5-3: Find $\lim\limits_{x \to \pi/2} (\sin x)^x$.
Using the property

$$\lim_{x \to a} f(x)^{g(x)} = \left(\lim_{x \to a} f(x) \right)^{\lim\limits_{x \to a} g(x)},$$

we have

$$\lim_{x \to \pi/2} (\sin x)^x = \left(\lim_{x \to \pi/2} \sin x \right)^{\lim\limits_{x \to \pi/2} x}.$$

Plugging in $x = \frac{\pi}{2}$,

$$\lim_{x \to \pi/2} \sin x = \sin\left(\frac{\pi}{2}\right) = 1,$$

and

$$\lim_{x \to \pi/2} x = \frac{\pi}{2}.$$

Thus, our final answer is

$$\lim_{x \to \pi/2} (\sin x)^x = 1^{\pi/2} = 1.$$

2.6 Squeeze Theorem

In this section, we will discuss the squeeze theorem, also known as "sandwich theorem." As the name of the theorem suggests, there are three functions involved in this theorem, and two functions are "squeezing" another function in the middle.

Squeeze Theorem (Sandwich Theorem)
Assume that functions $f(x)$, $g(x)$, and $h(x)$ defined in $D \subseteq \mathbb{R}$ satisfy

$$g(x) \leq f(x) \leq h(x), \forall x \in D.$$

Then, for some a in D, if $\lim_{x \to a} g(x) = \lim_{x \to a} h(x) = L$, then $\lim_{x \to a} f(x) = L$.

The notations might be a little confusing since you are not expected to be familiar with set notations before learning calculus. $\mathbb{R}$ denotes the set of real numbers. "$D \subseteq \mathbb{R}$" means "D is subset of $\mathbb{R}$". A set D is subset of $\mathbb{R}$ when all elements in set D are in $\mathbb{R}$. "$\forall x \in D$" means "for all x in set D".

If you think that this theorem is hard to remember, think of a sandwich. $g(x)$ and $h(x)$ are two pieces of bread, and $f(x)$ is a piece of fresh ham in between the two pieces of bread. If you are a vegetarian or a vegan, imagine $f(x)$ as fresh tomatoes in the sandwich.

Now let us take a look at an example using squeeze theorem.

Example 2.6-1: Find $\lim_{x \to 0} \frac{\sin x}{x}$ using squeeze theorem.
First, we need to derive an inequality that can be used to apply the

squeeze theorem. Now let us take look at the quadrant I of the unit circle.

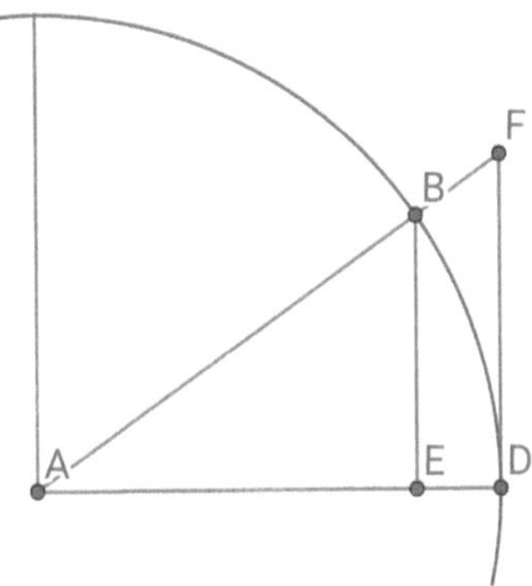

Figure 2.5: Quadrant I of unit circle

In figure 2.5, we have AD $= 1$, $\overline{\text{BE}} \perp \overline{\text{AD}}$, and $\overline{\text{FD}} \perp \overline{\text{AD}}$. Let

$$m\angle\text{BAD} = x,$$

where x is measured in radians. Then, we have BE $= \sin x$ and AE $= \cos x$.

Since $m\angle\text{BAE} = m\angle\text{FAD} = x$ and $m\angle\text{BEA} = m\angle\text{FDA} = \frac{\pi}{2}$,

$$\triangle\text{ABE} \sim \triangle\text{AFD}$$

by angle-angle similarity. Since $\triangle\text{ABE} \sim \triangle\text{AFD}$, we have

$$\begin{aligned} \frac{\text{BE}}{\text{AE}} &= \frac{\text{FD}}{\text{AD}} \\ \frac{\sin x}{\cos x} &= \frac{\text{FD}}{1} \\ \tan x &= \text{FD}. \end{aligned}$$

Now let A_1 be the area of $\triangle\text{ABD}$, A_2 be the area of sector ABD, and A_3 be the area of $\triangle\text{AFD}$. We can see that in quadrant I of the unit circle

$$A_1 \le A_2 \le A_3.$$

We have

$$A_1 = \text{AB} \times \text{AD} \times \sin\left(m\angle\text{BAD}\right) \times \frac{1}{2} = \frac{\sin x}{2},$$

$$A_2 = \pi \times \mathrm{AD}^2 \times \frac{m\angle BAD}{2\pi} = \frac{x}{2},$$

and

$$A_3 = \mathrm{FD} \times \mathrm{AD} \times \frac{1}{2} = \frac{\tan x}{2}.$$

Thus,

$$\frac{\sin x}{2} \leq \frac{x}{2} \leq \frac{\tan x}{2}. \tag{2.3}$$

Multiplying (2.3) by 2,

$$\sin x \leq x \leq \tan x. \tag{2.4}$$

Dividing (2.4) by $\sin x$,

$$1 \leq \frac{x}{\sin x} \leq \frac{1}{\cos x}. \tag{2.5}$$

The inequality signs did not change because $\sin x$ is positive for $x \in \left(0, \frac{\pi}{2}\right)$. Taking the reciprocals of (2.5),

$$\cos x \leq \frac{\sin x}{x} \leq 1 \tag{2.6}$$

for $x \in \left(0, \frac{\pi}{2}\right)$. Since $\cos x$ and $\frac{\sin x}{x}$ are even functions, the inequality in (2.6) also holds for $x \in \left(-\frac{\pi}{2}, 0\right)$. Thus,

$$\cos x \leq \frac{\sin x}{x} \leq 1$$

for any non-zero x on the interval $\left(-\frac{\pi}{2}, \frac{\pi}{2}\right)$. Note that

$$\lim_{x \to 0} 1 = 1.$$

Plugging in $x = 0$,

$$\lim_{x \to 0} \cos x = \cos 0 = 1.$$

By squeeze theorem, our final answer is

$$\lim_{x \to 0} \frac{\sin x}{x} = 1.$$

Because $\frac{\sin x}{x}$ is undefined at $x = 0$, we can say that $\frac{\sin x}{x}$ has removable discontinuity at the point (0,1).

From figure 2.6, we can see that indeed

$$\lim_{x \to 0} \frac{\sin x}{x} = 1.$$

There should be a hole at (0,1), but Desmos does not show the hole unless you hold the mouse above that point.

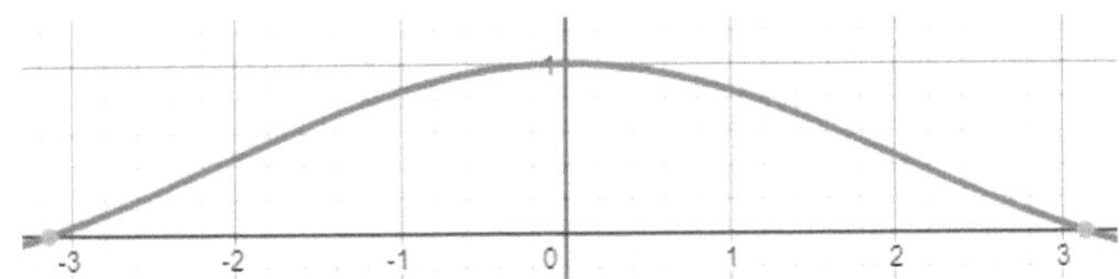

Figure 2.6: Graph of $f(x) = \frac{\sin x}{x}$

Example 2.6-2: Find $\lim_{x \to 0} \dfrac{x \sin x}{1+x^2}$ using squeeze theorem.
Although we can evaluate this limit by simply plugging $x = 0$ into the expression

$$\frac{x \sin x}{1+x^2},$$

I will show you the way to evaluate this limit using squeeze theorem, so that you will know how to handle similar problems that are more complicated.

Since

$$-1 \leq \sin x \leq 1$$

for $x \in \mathbb{R}$, the function

$$f(x) = \frac{x \sin x}{1+x^2}$$

is "squeezed" by two functions

$$g(x) = \frac{-x}{1+x^2}$$

and

$$h(x) = \frac{x}{1+x^2}.$$

Plugging $x = 0$ yields

$$\lim_{x \to 0} \frac{-x}{1+x^2} = \frac{0}{1+0^2} = 0 \tag{2.7}$$

and

$$\lim_{x \to 0} \frac{x}{1+x^2} = \frac{0}{1+0^2} = 0. \tag{2.8}$$

Therefore, from (2.7) and (2.8) and by squeeze theorem, we obtain

$$\lim_{x \to 0} \frac{x \sin x}{1+x^2} = 0.$$

Exercise Problems

1. Given that $\lim_{x \to \infty} \sqrt[x]{x} = 1$, find $\lim_{x \to \infty} \left(\sqrt[x]{x} + \frac{6x^3 + 5x + 1}{2x^3 + 4x^2 - 8} \right)$.

2. For any integer n, simplify the following expressions:
   a) $\lim_{x \to n\pi} \sin x$,
   b) $\lim_{x \to n\pi} \cos x$.

3. What two functions "squeeze" the function $f(x) = x^3 \cos x$?

4. By observing the graph of $f(x) = e^{-x}$, find the value of $\lim_{x \to \infty} e^{-x}$.

5. Evaluate $\lim_{x \to \infty} \frac{x + \cos x}{x}$. (Hint: use squeeze theorem)

Chapter 3

Introduction to Ordinary Derivatives

3.1 Introduction

In this chapter, we will learn about derivatives, or more specifically ordinary derivatives. There is another type of derivatives called partial derivatives, which we will discuss in chapter 14. An ordinary derivative is the derivative of a single variable function with respect to the variable of that function. A partial derivative is the derivative of a multivariable function with respect to one of the variables of that function.

From this point on in this book, when I say "derivatives", I am referring to ordinary derivatives. In chapter 14, when we discuss partial derivatives, I will refer to partial derivatives by specifically saying "partial derivatives".

3.2 Definition of Derivatives

In this section, I would like to introduce the definition of derivative as a limit and explain what the derivative of a function is in an intuitive way,

Definition of Derivative
Let $f(x)$ be a function in terms of x. Then, the derivative of $f(x)$ with respect to x, denoted by $f'(x)$ or $\frac{d}{dx}(f(x))$, is defined as

$$\frac{d}{dx}(f(x)) = f'(x) = \lim_{h \to 0} \frac{f(x+h) - f(x)}{h}.$$

Note that

$$\frac{f(x+h) - f(x)}{h}$$

is the slope of the secant line through the points $(x, f(x))$ and $(x+h, f(x+h))$.

As h approaches 0, the distance between the two points $(x, f(x))$ and $(x+h, f(x+h))$ becomes infinitesimal, meaning infinitely small, so

$$\lim_{h \to 0} \frac{f(x+h) - f(x)}{h}$$

becomes the slope of the tangent line at the point $(x, f(x))$.

A secant line is a line that crosses a curve at two or more points, and a tangent line is a line that barely touches a curve at some point.

Thus, graphically, the derivative of a function $f(x)$ at a certain point, say x_0, is the slope of the tangent line to the curve of $f(x)$ at the point $(x_0, f(x_0))$.

Below is the illustration of $f'(1)$ for $f(x) = x^2$.

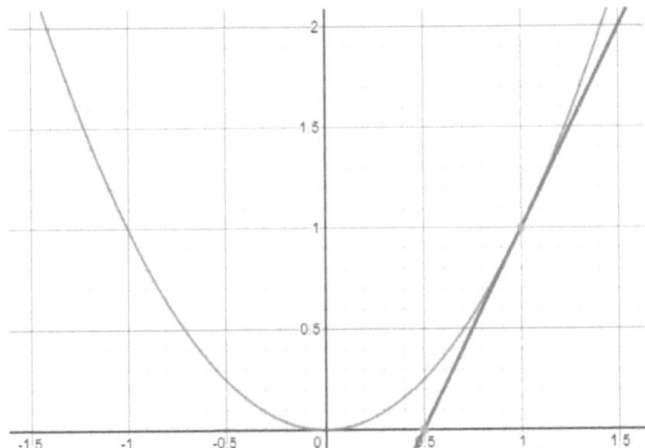

Figure 3.1: For $f(x) = x^2$, $f'(1)$ is the slope of the tangent line to the curve $y = x^2$ at the point $x = 1$.

Here are a few examples of problems using the definition of derivative.

Example 3.2-1: Prove that $\frac{d}{dx}\left(x^2\right) = 2x$.
Using the definition of derivative,

$$\frac{d}{dx}\left(x^2\right) = \lim_{h\to 0}\frac{(x+h)^2 - x^2}{h}. \tag{3.1}$$

Evaluating the limit on the RHS of (3.1),

$$\begin{aligned}\lim_{h\to 0}\frac{(x+h)^2 - x^2}{h} &= \lim_{h\to 0}\frac{x^2 + 2hx + h^2 - x^2}{h} \\ &= \lim_{h\to 0}(2x+h) \\ &= 2x.\end{aligned}$$

Therefore,

$$\frac{d}{dx}\left(x^2\right) = 2x.$$

Before we start the next example, I would like to formally introduce the Napier's constant, e, which we have used many times throughout chapter 1.

Napier's constant
Napier's constant, e, is defined as

$$e = \lim_{n\to\infty}\left(1+\frac{1}{n}\right)^n = \lim_{x\to 0}(1+x)^{\frac{1}{x}}.$$

More generally, we have the formula

$$\lim_{n\to\infty}\left(1+\frac{a}{n}\right)^{bn} = e^{ab}. \tag{3.2}$$

Example 3.2-2: Prove that $\frac{d}{dx}(e^x) = e^x$.
Using the definition of derivative,

$$\frac{d}{dx}(e^x) = \lim_{h\to 0}\frac{e^{x+h}-e^x}{h}. \tag{3.3}$$

Evaluating the limit on the RHS of (3.3),

$$\begin{aligned}\lim_{h\to 0}\frac{e^{x+h}-e^x}{h} &= \lim_{h\to 0}\frac{e^x e^h - e^x}{h}\\ &= \lim_{h\to 0}\frac{e^x\left(e^h-1\right)}{h}\\ &= e^x\lim_{h\to 0}\frac{\left(e^h-1\right)}{h}.\end{aligned}$$

In the last step, we take e^x out of the limit because e^x is independent of h. Let us substitute $u = e^h - 1$, or equivalently $h = \ln(u+1)$. Since

$$\lim_{h\to 0}\left(e^h-1\right) = e^0 - 1 = 0,$$

u approaches 0 as h approaches 0. After the substitution, our limit becomes

$$\begin{aligned}\lim_{h\to 0}\frac{\left(e^h-1\right)}{h} &= \lim_{u\to 0}\frac{u}{\ln(u+1)}\\ &= \lim_{u\to 0}\frac{1}{\left(\frac{\ln(u+1)}{u}\right)}\\ &= \lim_{u\to 0}\frac{1}{\ln\left((u+1)^{\frac{1}{u}}\right)}.\end{aligned}$$

By definition,

$$\lim_{u\to 0}(u+1)^{\frac{1}{u}} = e,$$

so

$$\lim_{u\to 0}\ln\left((u+1)^{\frac{1}{u}}\right)=\ln e=1.$$

Thus, our limit is

$$\lim_{h\to 0}\frac{(e^h-1)}{h}=\lim_{u\to 0}\frac{1}{\ln\left((u+1)^{\frac{1}{u}}\right)}=\frac{1}{1}=1.$$

Plugging this value back into the original limit of (3.3),

$$\lim_{h\to 0}\frac{e^{x+h}-e^x}{h}=e^x\lim_{h\to 0}\frac{(e^h-1)}{h}=e^x.$$

Therefore,

$$\frac{d}{dx}\left(e^x\right)=e^x.$$

Example 3.2-3: Prove that $\frac{d}{dx}(\ln x)=\frac{1}{x}$.
Using the definition of derivative,

$$\frac{d}{dx}(\ln x)=\lim_{h\to 0}\frac{\ln(x+h)-\ln x}{h}. \tag{3.4}$$

Now let us break down the two-sided limit in (3.4) into two one-sided limits:

$$\lim_{h\to 0^+}\frac{\ln(x+h)-\ln x}{h} \tag{3.5}$$

and

$$\lim_{h\to 0^-}\frac{\ln(x+h)-\ln x}{h}. \tag{3.6}$$

First, let us evaluate the limit in (3.5), which is the limit as h approaches 0 from the right. Using the property of logarithms,

$$\lim_{h\to 0^+}\frac{\ln(x+h)-\ln x}{h}=\lim_{h\to 0^+}\frac{\ln\left(\frac{x+h}{x}\right)}{h}=\lim_{h\to 0^+}\frac{\ln\left(1+\frac{h}{x}\right)}{h}.$$

Now let us substitute $u=\frac{1}{h}$, or equivalently $h=\frac{1}{u}$. Since

$$\lim_{h\to 0^+}\frac{1}{h}=\infty,$$

u approaches ∞ as h approaches 0 from the right (or 0^+). After the substitution, our limit becomes

$$\lim_{h\to 0^+}\frac{\ln\left(1+\frac{h}{x}\right)}{h}=\lim_{u\to\infty}u\ln\left(1+\frac{1}{ux}\right)=\lim_{u\to\infty}\ln\left(\left(1+\frac{1}{ux}\right)^u\right).$$

Using the formula in (3.2),

$$\lim_{u\to\infty}\left(1+\frac{1}{ux}\right)^u = e^{\frac{1}{x}},$$

so

$$\lim_{u\to\infty}\ln\left(\left(1+\frac{1}{ux}\right)^u\right) = \ln\left(e^{\frac{1}{x}}\right) = \frac{1}{x}.$$

Thus, the limit in (3.5) is

$$\lim_{h\to 0^+}\frac{\ln(x+h)-\ln x}{h} = \frac{1}{x}. \tag{3.7}$$

Next, let us evaluate the limit in (3.6), which is the limit as h approaches 0 from the left. Using the property of logarithms,

$$\lim_{h\to 0^-}\frac{\ln(x+h)-\ln x}{h} = \lim_{h\to 0^-}\frac{\ln\left(\frac{x+h}{x}\right)}{h} = \lim_{h\to 0^-}\frac{\ln\left(1+\frac{h}{x}\right)}{h}.$$

Now let us substitute $u=-\frac{1}{h}$, or equivalently $h=-\frac{1}{u}$. Since

$$\lim_{h\to 0^-}-\frac{1}{h} = \infty,$$

u approaches ∞ as h approaches 0 from the left (or 0^-). If you do not understand why

$$\lim_{h\to 0^-}-\frac{1}{h} = \infty,$$

think of the graph of $f(h)=-\frac{1}{h}$. After the substitution, our limit becomes

$$\lim_{h\to 0^-}\frac{\ln\left(1+\frac{h}{x}\right)}{h} = \lim_{u\to\infty}-u\ln\left(1-\frac{1}{ux}\right) = \lim_{u\to\infty}\ln\left(\left(1-\frac{1}{ux}\right)^{-u}\right).$$

Using the formula in (3.2),

$$\lim_{u\to\infty}\left(1-\frac{1}{ux}\right)^{-u} = e^{\frac{1}{x}},$$

so

$$\lim_{u\to\infty}\ln\left(\left(1-\frac{1}{ux}\right)^{-u}\right) = \ln\left(e^{\frac{1}{x}}\right) = \frac{1}{x}.$$

Thus, the limit in (3.6) is

$$\lim_{h\to 0^-}\frac{\ln(x+h)-\ln x}{h} = \frac{1}{x}. \tag{3.8}$$

From (3.7) and (3.8), we obtain

$$\lim_{h\to 0^-}\frac{\ln(x+h)-\ln x}{h}=\lim_{h\to 0^+}\frac{\ln(x+h)-\ln x}{h}=\frac{1}{x}.$$

Since both one-sided limits agree, the two-sided limit in (3.4) exists, and

$$\lim_{h\to 0}\frac{\ln(x+h)-\ln x}{h}=\frac{1}{x}.$$

Therefore,

$$\frac{d}{dx}(\ln x)=\frac{1}{x}.$$

In this chapter, we have evaluated derivatives using the limit definition of derivative. However, later on, we will not have to use the limit definition of derivative anymore. There are neat formulae and rules for derivatives, so we just need to simply apply those formulae and rules, which will help us evaluate derivatives faster and more easily.

3.3 Basic Properties of Derivatives

These are the basic properties of derivatives that will be useful for evaluating derivatives.

Basic Properties of Derivatives
Let $f(x)$ and $g(x)$ be functions in terms of x and k be a constant.

$$\frac{d}{dx}(kf(x))=k\frac{d}{dx}(f(x))$$
$$\frac{d}{dx}(f(x)\pm g(x))=\frac{d}{dx}(f(x))\pm\frac{d}{dx}(g(x))$$

We will use these properties in many problems throughout the rest of this chapter and the rest of this book.

3.4 Power Rule

In this section, we will discuss the power rule and a few examples using the power rule.

Power Rule
For n any complex number,

$$\frac{d}{dx}\left(x^n\right) = nx^{n-1}.$$

Example 3.4-1: Evaluate $\frac{d}{dx}\left(\sqrt{x}\right)$.
Note that

$$\sqrt{x} = x^{\frac{1}{2}}.$$

Thus,

$$\frac{d}{dx}\left(\sqrt{x}\right) = \frac{d}{dx}\left(x^{\frac{1}{2}}\right).$$

Applying the power rule,

$$\frac{d}{dx}\left(x^{\frac{1}{2}}\right) = \frac{1}{2}x^{\frac{1}{2}-1} = \frac{1}{2\sqrt{x}}.$$

Therefore, our final answer is

$$\frac{d}{dx}\left(\sqrt{x}\right) = \frac{1}{2\sqrt{x}}.$$

Example 3.4-2: Evaluate $\frac{d}{dx}\left(x^5 + x^2 - \frac{1}{x}\right)$
Using a property we just learned in the section 3.3,

$$\begin{aligned}\frac{d}{dx}\left(x^5 + x^2 - \frac{1}{x}\right) &= \frac{d}{dx}\left(x^5\right) + \frac{d}{dx}\left(x^2\right) - \frac{d}{dx}\left(\frac{1}{x}\right)\\ &= \frac{d}{dx}\left(x^5\right) + \frac{d}{dx}\left(x^2\right) - \frac{d}{dx}\left(x^{-1}\right).\end{aligned}$$

Applying power rule,

$$\begin{aligned}\frac{d}{dx}\left(x^5\right) + \frac{d}{dx}\left(x^2\right) - \frac{d}{dx}\left(x^{-1}\right) &= 5x^4 + 2x^1 - \left(-x^{-2}\right)\\ &= 5x^4 + 2x + \frac{1}{x^2}.\end{aligned}$$

Therefore,

$$\frac{d}{dx}\left(x^5 + x^2 - \frac{1}{x}\right) = 5x^4 + 2x + \frac{1}{x^2}.$$

3.5 Derivatives of Exponential Functions

In section 3.2, we have looked at the derivative of the exponential function when the base is e, $\frac{d}{dx}\left(e^x\right)$. In this section, we will look at the general case when the base is any complex number a.

Derivatives of Exponential Functions
For a any non-zero complex number,

$$\frac{d}{dx}\left(a^x\right) = a^x \ln a.$$

You might say that a cannot be any complex number because the natural logarithmic function is only defined for positive real numbers, but the natural logarithmic function can actually be extended to negative real numbers and even complex numbers.

For example, $\ln(-1) = i\pi$, where $i = \sqrt{-1}$, if we only use the principal branch. There are many more values for $\ln(-1)$ other than $i\pi$. However, I will not go into complex logarithms in detail because it is beyond the scope of this book.

Note that when $a = e$, we would have

$$\frac{d}{dx}\left(e^x\right) = e^x \ln e = e^x,$$

which is the result we got in example 3.2-2.

Example 3.5-1: Let $f(x) = 3^x + 2^{-x}$. Find $f'(2)$.
First, we have

$$f'(x) = \frac{d}{dx}\left(3^x + 2^{-x}\right) = \frac{d}{dx}\left(3^x\right) + \frac{d}{dx}\left(2^{-x}\right).$$

Since $2^{-x} = \left(2^{-1}\right)^x$,

$$\frac{d}{dx}\left(3^x\right) + \frac{d}{dx}\left(2^{-x}\right) = \frac{d}{dx}\left(3^x\right) + \frac{d}{dx}\left(\left(2^{-1}\right)^x\right) = 3^x \ln 3 + \left(2^{-1}\right)^x \ln\left(2^{-1}\right).$$

Using the property of logarithms, $\ln\left(2^{-1}\right) = -\ln 2$, so

$$3^x \ln 3 + \left(2^{-1}\right)^x \ln\left(2^{-1}\right) = 3^x \ln 3 - 2^{-x} \ln 2.$$

Plugging in $x = 2$, our final answer is

$$f'(2) = 3^2 \ln 3 - 2^{-2} \ln 2 = 9 \ln 3 - \frac{\ln 2}{4}.$$

3.6 Derivatives of Logarithmic Functions

In section 3.2, we have looked at the derivative of the logarithmic function when the base is e, or the natural logarithm, $\frac{d}{dx}(\ln x)$. In this section, we will look at the general case when the base is, again, any complex number a.

Derivatives of Logarithmic Functions
For a any non-zero complex number,

$$\frac{d}{dx}(\log_a x) = \frac{1}{x \ln a}.$$

Note that when $a = e$, we would have

$$\frac{d}{dx}(\ln x) = \frac{1}{x \ln e} = \frac{1}{x},$$

which is the result we got in example 3.2-3.

Example 3.6-1: Let $f(x) = 3\log_4 x$. Find $f'(5)$.
First, we have

$$f'(x) = \frac{d}{dx}(3\log_4 x) = 3\frac{d}{dx}(\log_4 x).$$

Using the formula for derivatives of logarithmic functions,

$$3\frac{d}{dx}(\log_4 x) = \frac{3}{x \ln 4}.$$

Plugging in $x = 5$, our final answer is

$$f'(5) = \frac{3}{5 \ln 4}.$$

3.7 Derivatives of Trigonometric Functions

In this section, we will discuss the derivatives of the six trigonometric functions.

Derivatives of Trigonometric Functions

$$\frac{d}{dx}(\sin x) = \cos x$$
$$\frac{d}{dx}(\cos x) = -\sin x$$
$$\frac{d}{dx}(\tan x) = \sec^2 x$$
$$\frac{d}{dx}(\cot x) = -\csc^2 x$$
$$\frac{d}{dx}(\sec x) = \sec x \tan x$$
$$\frac{d}{dx}(\csc x) = -\csc x \cot x$$

In chapter 13, I will prove the derivatives of $\sin x$ and $\cos x$ using the definitions of $\sin x$ and $\cos x$ derived from Euler's formula. Once the derivatives of $\sin x$ and $\cos x$ are proven, we can prove the derivatives of other trigonometric functions using the quotient rule, which we will discuss in chapter 4.

Example 3.7-1: Let $f(x) = \tan x + 2\sec x$. Find $f'\left(\frac{\pi}{4}\right)$.
First, we have

$$f'(x) = \frac{d}{dx}(\tan x + 2\sec x) = \frac{d}{dx}(\tan x) + 2\frac{d}{dx}(\sec x).$$

Since

$$\frac{d}{dx}(\tan x) = \sec^2 x$$

and

$$\frac{d}{dx}(\sec x) = \sec x \tan x,$$

we obtain

$$f'(x) = \sec^2 x + 2\sec x \tan x.$$

Plugging in $x = \frac{\pi}{4}$, our final answer is

$$f'\left(\frac{\pi}{4}\right) = \sec^2\left(\frac{\pi}{4}\right) + 2\sec\left(\frac{\pi}{4}\right)\tan\left(\frac{\pi}{4}\right) = 2 + 2\sqrt{2}.$$

3.8 Derivatives of Inverse Trigonometric Functions

Here are the derivatives of inverse trigonometric functions.

Derivatives of Inverse Trigonometric Functions

$$\frac{d}{dx}(\arcsin x) = \frac{1}{\sqrt{1-x^2}}$$
$$\frac{d}{dx}(\arccos x) = \frac{-1}{\sqrt{1-x^2}}$$
$$\frac{d}{dx}(\arctan x) = \frac{1}{1+x^2}$$
$$\frac{d}{dx}(\text{arccot}\, x) = \frac{-1}{1+x^2}$$
$$\frac{d}{dx}(\text{arcsec}\, x) = \frac{1}{|x|\sqrt{x^2-1}}$$
$$\frac{d}{dx}(\text{arccsc}\, x) = \frac{-1}{|x|\sqrt{x^2-1}}$$

Example 3.8-1: Let $f(x) = \arctan x$. Find $f'(1)$.
First, we have

$$f'(x) = \frac{d}{dx}(\arctan x) = \frac{1}{1+x^2}.$$

Plugging in $x = 1$, our final answer is

$$f'(1) = \frac{1}{1+1^2} = \frac{1}{2}.$$

3.9 Derivatives of Hyperbolic Functions

In this section, we will discuss the derivatives of the six hyperbolic functions.

Derivatives of Hyperbolic Functions

$$\frac{d}{dx}(\sinh x) = \cosh x$$
$$\frac{d}{dx}(\cosh x) = \sinh x$$
$$\frac{d}{dx}(\tanh x) = \operatorname{sech}^2 x$$
$$\frac{d}{dx}(\coth x) = -\operatorname{csch}^2 x$$
$$\frac{d}{dx}(\operatorname{sech} x) = -\operatorname{sech} x \tanh x$$
$$\frac{d}{dx}(\operatorname{csch} x) = -\operatorname{csch} x \coth x$$

In the following examples, I will prove the derivatives of $\sinh x$ and $\cosh x$ using the definitions of $\sinh x$ and $\cosh x$, which we learned in chapter 1. Once the derivatives of $\sinh x$ and $\cosh x$ are proven, we can prove the derivatives of the other four hyperbolic functions using the quotient rule, which we will discuss in chapter 4.

Example 3.9-1: Prove that $\frac{d}{dx}(\sinh x) = \cosh x$.
By definition,

$$\sinh x = \frac{e^x - e^{-x}}{2}.$$

Thus,

$$\frac{d}{dx}(\sinh x) = \frac{d}{dx}\left(\frac{e^x - e^{-x}}{2}\right).$$

Using the properties of derivatives that we learned in section 3.3,

$$\frac{d}{dx}\left(\frac{e^x - e^{-x}}{2}\right) = \frac{1}{2}\frac{d}{dx}\left(e^x\right) - \frac{1}{2}\frac{d}{dx}\left(e^{-x}\right).$$

Applying the formula for derivatives of exponential functions,

$$\frac{d}{dx}\left(e^x\right) = e^x \ln \mathrm{e} = e^x,$$

and

$$\frac{d}{dx}\left(e^{-x}\right) = e^{-x} \ln\left(e^{-1}\right) = -e^{-x}.$$

So,

$$\frac{1}{2}\frac{d}{dx}\left(e^x\right) - \frac{1}{2}\frac{d}{dx}\left(e^{-x}\right) = \frac{1}{2}e^x - \frac{1}{2}\left(-e^{-x}\right) = \frac{e^x + e^{-x}}{2}.$$

By definition,

$$\frac{e^x + e^{-x}}{2} = \cosh x.$$

Therefore,

$$\frac{d}{dx}(\sinh x) = \cosh x.$$

Example 3.9-2: Prove that $\frac{d}{dx}(\cosh x) = \sinh x$.
By definition,

$$\cosh x = \frac{e^x + e^{-x}}{2}.$$

Thus,

$$\frac{d}{dx}(\cosh x) = \frac{d}{dx}\left(\frac{e^x + e^{-x}}{2}\right).$$

Using the properties of derivatives that we learned in section 3.3,

$$\frac{d}{dx}\left(\frac{e^x + e^{-x}}{2}\right) = \frac{1}{2}\frac{d}{dx}\left(e^x\right) + \frac{1}{2}\frac{d}{dx}\left(e^{-x}\right).$$

Applying the formula for derivatives of exponential functions,

$$\frac{d}{dx}\left(e^x\right) = e^x \ln e = e^x,$$

and

$$\frac{d}{dx}\left(e^{-x}\right) = e^{-x} \ln\left(e^{-1}\right) = -e^{-x}.$$

So,

$$\frac{1}{2}\frac{d}{dx}\left(e^x\right) + \frac{1}{2}\frac{d}{dx}\left(e^{-x}\right) = \frac{1}{2}e^x + \frac{1}{2}\left(-e^{-x}\right) = \frac{e^x - e^{-x}}{2}.$$

By definition,

$$\frac{e^x - e^{-x}}{2} = \sinh x.$$

Therefore,

$$\frac{d}{dx}(\cosh x) = \sinh x.$$

3.10 Derivatives of Inverse Hyperbolic Functions

In this section, we will discuss the derivatives of inverse hyperbolic functions.

Derivatives of Inverse Hyperbolic Functions

$$\frac{d}{dx}\left(\sinh^{-1} x\right) = \frac{1}{\sqrt{1+x^2}}$$
$$\frac{d}{dx}\left(\cosh^{-1} x\right) = \frac{1}{\sqrt{x^2-1}}$$
$$\frac{d}{dx}\left(\tanh^{-1} x\right) = \frac{1}{1-x^2}$$
$$\frac{d}{dx}\left(\coth^{-1} x\right) = \frac{1}{1-x^2}$$
$$\frac{d}{dx}\left(\operatorname{sech}^{-1} x\right) = \frac{-1}{x\sqrt{1-x^2}}$$
$$\frac{d}{dx}\left(\operatorname{csch}^{-1} x\right) = \frac{-1}{|x|\sqrt{1+x^2}}$$

We can see that the derivatives of $\tanh^{-1} x$ and $\coth^{-1} x$ are the same. However, the domains of $\tanh^{-1} x$ and $\coth^{-1} x$ are different. The domain of $\tanh^{-1} x$ is $(-1,1)$ or $\{x \mid |x| < 1\}$ while the domain of $\coth^{-1} x$ is $(-\infty,-1) \cup (1,\infty)$ or $\{x \mid |x| > 1\}$.

Example 3.10-1: Let $f(x) = \coth^{-1} x$. Find $f'(2)$.
First, we have

$$f'(x) = \frac{d}{dx}\left(\coth^{-1} x\right) = \frac{1}{1-x^2}.$$

Plugging in $x = 2$, our final answer is

$$f'(2) = \frac{1}{1-2^2} = \frac{-1}{3}.$$

3.11 Derivative of the Absolute Value Function

In this section, we will discuss the derivative of the absolute value function. First, we will define the derivative of the absolute value function as a piece-wise function. Then I will introduce the formula for the derivative of the absolute value function without using the piece-wise definition.

First, let us take a look at the graph of $y = |x|$ in figure 3.2.

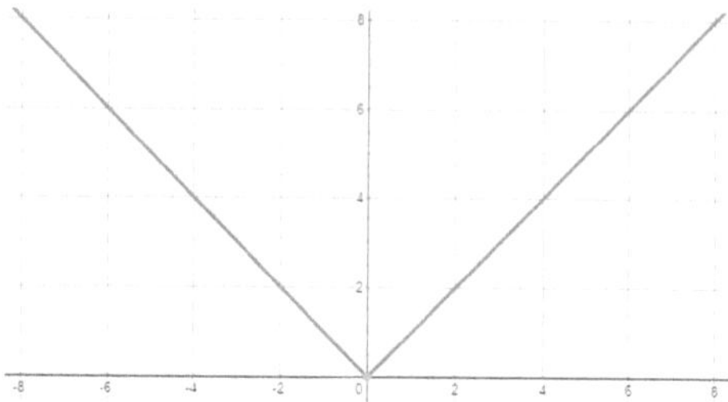

Figure 3.2: Graph of $y = |x|$

We can see that for $x > 0$, the graph of $y = |x|$ looks like that of $y = x$. For $x < 0$, the graph of $y = |x|$ looks like that of $y = -x$. The observations above are true since, by definition, $|x| = x$ for $x > 0$ and $|x| = -x$ for $x < 0$.

Recall that the derivative of a function at a certain point is the slope of that function at that point. For linear functions like $y = x$ or $y = -x$, the slopes are constant, which means they do not change. The slope of $y = x$ is 1, and the slope of $y = -x$ is -1.

Thus,

$$\frac{d}{dx}(|x|) = \frac{d}{dx}(x) = 1$$

for $x > 0$, and

$$\frac{d}{dx}(|x|) = \frac{d}{dx}(-x) = -1$$

for $x < 0$.

Also, note that we can find the derivatives of x and $-x$ using the power rule.

At the point $x = 0$, the derivative is undefined since it is a sharp point. If you do not understand what a sharp point is, do not overthink it. A sharp point is literally a "sharp" point. Imagine if you have an object of the shape of the graph of $y = |x|$ and the point at $x = 0$ hits you hard, you are going to need to go to the nurse.

Also, recall that the definition of derivative is

$$f'(x) = \lim_{h \to 0} \frac{f(x+h) - f(x)}{h}.$$

Let $f(x) = |x|$ and substituting $x = 0$,

$$f'(0) = \lim_{h \to 0} \frac{|h|}{h}.$$

To prove that the derivative is undefined at $x = 0$, we need to prove that the limit above does not exist. Let us first take a look at the limit as h approaches 0 from the right:

$$\lim_{h \to 0+} \frac{|h|}{h}.$$

When h is approaching 0 from the right, h is positive. So, $|h| = h$, and we have

$$\lim_{h \to 0+} \frac{|h|}{h} = \lim_{h \to 0+} \frac{h}{h} = 1. \tag{3.9}$$

Next, let us take a look at the limit as h approaches 0 from the left:

$$\lim_{h \to 0-} \frac{|h|}{h}.$$

When h is approaching 0 from the left, h is negative. So, $|h| = -h$, and we have

$$\lim_{h \to 0-} \frac{|h|}{h} = \lim_{h \to 0-} \frac{-h}{h} = -1. \tag{3.10}$$

From (3.9) and (3.10), we can see that the two one-sided limits do not agree:

$$\lim_{h \to 0+} \frac{|h|}{h} \neq \lim_{h \to 0-} \frac{|h|}{h}.$$

Thus, the two-sided limit

$$\lim_{h \to 0} \frac{|h|}{h}$$

does not exist, and the derivative of the function $f(x) = |x|$ is undefined at $x = 0$.

Hence, we can have the following piece-wise definition for $\frac{d}{dx}(|x|)$.

Piece-wise Definition for $\frac{\mathbf{d}}{\mathbf{dx}}(|\mathbf{x}|)$

$$\frac{d}{dx}(|x|) = \begin{cases} -1 & \text{if } x < 0 \\ \text{undefined} & \text{if } x = 0 \\ 1 & \text{if } x > 0 \end{cases}$$

However, there is a neat formula for the derivative of the absolute value function without using the piece-wise definition.

Derivative of the Absolute Value Function

$$\frac{d}{dx}(|x|) = \frac{x}{|x|}$$

Note that when $x < 0$,

$$\frac{d}{dx}(|x|) = \frac{x}{|x|} = \frac{x}{-x} = -1$$

because $|x| = -x$ when $x < 0$ by definition. Also, when $x > 0$,

$$\frac{d}{dx}(|x|) = \frac{x}{|x|} = \frac{x}{x} = 1$$

because $|x| = x$ when $x > 0$ by definition. When $x = 0$, the expression

$$\frac{x}{|x|}$$

is undefined. When the derivative of a function is undefined at a point, we say that the function is not differentiable at that point. Hence, this formula coincides with the piece-wise definition we derived on the last page.

3.12 Differentiability

We have looked at derivatives of many types of function. Now let us discuss differentiability of a function. Differentiability is the ability to differentiate, or take derivative.

If the derivative of a function is defined as the slope of the tangent line to the curve of the function, then the derivative is only defined where the function is continuous. A function is not differentiable at the points where the function is discontinuous. So, differentiability implies continuity.

However, does continuity imply differentiability? The answer is no. As we saw in the absolute value function, $|x|$ is not differentiable at $x = 0$ although it is continuous at $x = 0$. The reason was because the point at $x = 0$ was a sharp point.

Then, is there possibly a function that is continuous everywhere but differentiable nowhere? Yes, that function just needs to have sharp points everywhere. One such function is the Weierstrass function.

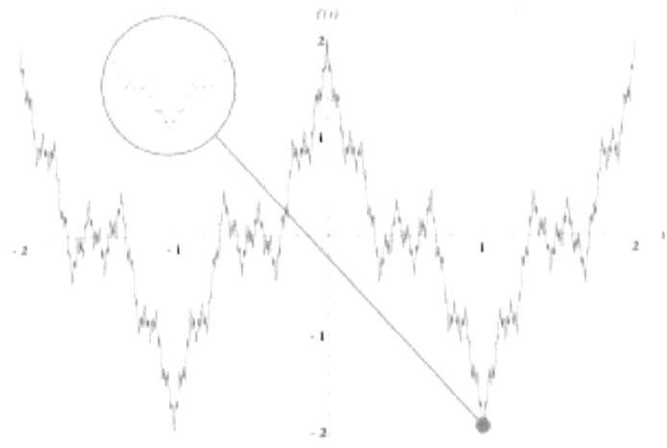

Figure 3.3: Graph of Weierstrass function on the interval $[-2, 2]$

Weierstrass Function
The Weierstrass function is defined as

$$f(x) = \sum_{n=0}^{\infty} a^n \cos\left(b^n \pi x\right)$$

where $0 < a < 1$, b is an odd positive integer, and $ab > 1 + \frac{3\pi}{2}$.

It is okay if you do not understand the $\sum$ notation. For now, you just need to know it is a function. The $\sum$ notation will be explained in chapter 11.

From figure 3.3, we can see that the Weierstrass function has sharp points everywhere, so it is differentiable nowhere. Also, it exhibits self-similarity. When we zoom in the function at a certain region, the behavior of the function at that region is similar to the behavior of the whole function, full of sharp points. Yeah, be careful not to get hit by the Weierstrass function: it is going to hurt much more than getting hit by the absolute value function at $x = 0$.

Exercise Problems

1. Using the definition of derivative, prove that $\frac{d}{dx}\left(x^3\right) = 3x^2$.

2. Using the definition of derivative, prove that $\frac{d}{dx}(\sin x) = \cos x$.

3. Evaluate $\frac{d}{dx}\left(\sqrt[3]{x}\right)$.

4. Evaluate $\frac{d}{dx}\left(5^{x\ln 5}\right)$.

5. Let $f(x) = \log_6 x$. Find the slope of the tangent line to the curve of $f(x)$ at $x = 1$.

6. Let $f(x) = g(x) + h(x)$ where

$$g(x) = \frac{4\sinh^2 x \cosh^2 x}{\cosh^4 x + 2\cosh^2 x \sinh^2 x + \sinh^4 x}$$

and

$$h(x) = \frac{\ln\left(x + \sqrt{x^2+1}\right)}{\sinh^{-1} x \left(1 + \cosh(4x) - \cosh^2(2x)\right)}.$$

Find $f'(x)$.

Chapter 4

Product Rule, Quotient Rule, and Chain Rule

4.1 Introduction

Product rule, quotient rule, and chain rule are three important rules in evaluating derivatives. Product rule is the rule for evaluating derivatives of a product of two or more functions. Quotient rule is the rule for evaluating derivatives of a quotient of two functions. Chain rule is the rule for evaluating derivatives of a composition of two or more functions.

In each of the following sections, we will discuss the rules one at a time. However, the chain rule that we discuss in this chapter is the chain rule for the two-dimensional xy-plane, which is the chain rule for ordinary derivatives of single variable functions. In chapter 14, I will introduce the generalized chain rule for partial derivatives of multivariable functions.

4.2 Product Rule

Here is the product rule.

Product Rule
For two functions $f(x)$ and $g(x)$ in terms of x,

$$\frac{d}{dx}(f(x)g(x)) = \frac{d}{dx}(f(x))g(x)+f(x)\frac{d}{dx}(g(x)) = f'(x)g(x)+f(x)g'(x).$$

We can also extend this to any n functions $f_1(x)$, $f_2(x)$, $f_3(x)$, $\cdots$, $f_n(x)$:

$$\begin{aligned}\frac{d}{dx}(f_1(x)f_2(x)f_3(x)\cdots f_n(x)) =& f_1'(x)f_2(x)f_3(x)\cdots f_n(x)\\ &+ f_1(x)f_2'(x)f_3(x)\cdots f_n(x)\\ &+ f_1(x)f_2(x)f_3'(x)\cdots f_n(x)\\ &+\cdots+ f_1(x)f_2(x)f_3(x)\cdots f_n'(x).\end{aligned}$$

Basically, for each term of the sum, we take the derivative of one of the functions each at a time, and the rest of the other functions stay the same. Let us take a look at some examples to get familiar with the product rule.

Example 4.2-1: Find the slope of the tangent line to the curve $y = x\sin x$ at $x = 1$.

Recall that the derivative of a function at a certain point is the slope of the tangent line to the curve of the function at that point. Thus, to solve this problem, we will take the derivative of y with respect to x, $\frac{dy}{dx}$, and plug in $x = 1$.

First, we have

$$\frac{dy}{dx} = \frac{d}{dx}(x \sin x).$$

Applying the product rule,

$$\frac{d}{dx}(x \sin x) = \frac{d}{dx}(x) \sin x + x \frac{d}{dx}(\sin x).$$

Since

$$\frac{d}{dx}(x) = 1$$

and

$$\frac{d}{dx}(\sin x) = \cos x,$$

we get

$$\frac{d}{dx}(x) \sin x + x \frac{d}{dx}(\sin x) = \sin x + x \cos x.$$

Plugging in $x = 1$, the slope of the tangent line to the curve $y = x \sin x$ at $x = 1$ is $\sin 1 + \cos 1$.

Example 4.2-2: Let $f(x) = x^2 e^x \sin x$. Evaluate $f'(x)$.
First, we have

$$f'(x) = \frac{d}{dx}\left(x^2 e^x \sin x\right).$$

Applying the product rule,

$$\frac{d}{dx}\left(x^2 e^x \sin x\right) = \frac{d}{dx}\left(x^2\right) e^x \sin x + x^2 \frac{d}{dx}\left(e^x\right) \sin x + x^2 e^x \frac{d}{dx}(\sin x).$$

Since

$$\frac{d}{dx}\left(x^2\right) = 2x,$$

$$\frac{d}{dx}\left(e^x\right) = e^x,$$

and

$$\frac{d}{dx}(\sin x) = \cos x,$$

we obtain

$$\begin{aligned} &\frac{d}{dx}\left(x^2\right) e^x \sin x + x^2 \frac{d}{dx}\left(e^x\right) \sin x + x^2 e^x \frac{d}{dx}(\sin x) \\ &= 2x e^x \sin x + x^2 e^x \sin x + x^2 e^x \cos x. \end{aligned}$$

Therefore, our final answer is

$$f'(x) = 2xe^x \sin x + x^2 e^x \sin x + x^2 e^x \cos x.$$

Example 4.2-3: Evaluate $\frac{d}{dx}(\cosh x \sinh x)$.
Applying the product rule,

$$\frac{d}{dx}(\cosh x \sinh x) = \frac{d}{dx}(\cosh x) \sinh x + \cosh x \frac{d}{dx}(\sinh x).$$

Since

$$\frac{d}{dx}(\cosh x) = \sinh x$$

and

$$\frac{d}{dx}(\sinh x) = \cosh x,$$

we obtain

$$\frac{d}{dx}(\cosh x) \sinh x + \cosh x \frac{d}{dx}(\sinh x) = \sinh^2 x + \cosh^2 x.$$

Since we have the identity

$$\sinh^2 x + \cosh^2 x = \cosh(2x),$$

our final answer is

$$\frac{d}{dx}(\cosh x \sinh x) = \cosh(2x).$$

Example 4.2-4: Let $f(x) = x|x| \ln x$. Find $f'(2)$.
First, we have

$$f'(x) = \frac{d}{dx}(x|x| \ln x).$$

Applying the product rule,

$$\frac{d}{dx}(x|x| \ln x) = \frac{d}{dx}(x)|x| \ln x + x\frac{d}{dx}(|x|) \ln x + x|x| \frac{d}{dx}(\ln x).$$

Since

$$\frac{d}{dx}(x) = \frac{d}{dx}\left(x^1\right) = 1x^{1-1} = 1,$$

$$\frac{d}{dx}(|x|) = \frac{x}{|x|},$$

and

$$\frac{d}{dx}(\ln x) = \frac{1}{x},$$

we have

$$\frac{d}{dx}(x)|x|\ln x + x\frac{d}{dx}(|x|)\ln x + x|x|\frac{d}{dx}(\ln x) = |x|\ln x + \frac{x^2\ln x}{|x|} + |x|$$
$$= \frac{|x|^2\ln x + x^2\ln x + |x|^2}{|x|}.$$

Since $|x|^2 = x^2$,

$$\frac{|x|^2\ln x + x^2\ln x + |x|^2}{|x|} = \frac{2x^2\ln x + x^2}{|x|}.$$

Thus,

$$f'(x) = \frac{2x^2\ln x + x^2}{|x|}.$$

Plugging in $x = 2$, our final answer is

$$f'(2) = 4\ln 2 + 2.$$

4.3 Quotient Rule

Here is the quotient rule.

Quotient Rule
For two functions $f(x)$ and $g(x)$ in terms of x,

$$\frac{d}{dx}\left(\frac{f(x)}{g(x)}\right) = \frac{f'(x)g(x) - f(x)g'(x)}{(g(x))^2}$$

if $g(x) \neq 0$.

Now let us get into a few examples to get familiar with the quotient rule. Also, as promised, I will show you how to prove the derivatives of the other four trigonometric functions other than $\sin x$ and $\cos x$ and the derivatives of the other four hyperbolic functions other than $\sinh x$ and $\cosh x$ using the quotient rule.

Example 4.3-1: Given that $\frac{d}{dx}(\sin x) = \cos x$ and $\frac{d}{dx}(\cos x) = -\sin x$, prove that $\frac{d}{dx}(\tan x) = \sec^2 x$.
By definition,

$$\tan x = \frac{\sin x}{\cos x}.$$

Thus,

$$\frac{d}{dx}(\tan x) = \frac{d}{dx}\left(\frac{\sin x}{\cos x}\right).$$

Applying the quotient rule,

$$\frac{d}{dx}\left(\frac{\sin x}{\cos x}\right) = \frac{\frac{d}{dx}(\sin x)\cos x - \sin x \frac{d}{dx}(\cos x)}{\cos^2 x}.$$

Since

$$\frac{d}{dx}(\sin x) = \cos x$$

and

$$\frac{d}{dx}(\cos x) = -\sin x,$$

we have

$$\frac{\frac{d}{dx}(\sin x)\cos x - \sin x \frac{d}{dx}(\cos x)}{\cos^2 x} = \frac{\cos^2 x + \sin^2 x}{\cos^2 x} = \frac{1}{\cos^2 x} = \sec^2 x.$$

Therefore,

$$\frac{d}{dx}(\tan x) = \sec^2 x.$$

Example 4.3-2: Given that $\frac{d}{dx}(\sin x) = \cos x$ and $\frac{d}{dx}(\cos x) = -\sin x$, prove that $\frac{d}{dx}(\cot x) = -\csc^2 x$.
By definition,

$$\cot x = \frac{\cos x}{\sin x}.$$

Thus,

$$\frac{d}{dx}(\cot x) = \frac{d}{dx}\left(\frac{\cos x}{\sin x}\right).$$

Applying the quotient rule,

$$\frac{d}{dx}\left(\frac{\cos x}{\sin x}\right) = \frac{\frac{d}{dx}(\cos x)\sin x - \cos x \frac{d}{dx}(\sin x)}{\sin^2 x}.$$

Since

$$\frac{d}{dx}(\sin x) = \cos x$$

and

$$\frac{d}{dx}(\cos x) = -\sin x,$$

we have

$$\frac{\frac{d}{dx}(\cos x)\sin x - \cos x \frac{d}{dx}(\sin x)}{\sin^2 x} = \frac{-\sin^2 x - \cos^2 x}{\sin^2 x} = \frac{-1}{\sin^2 x} = -\csc^2 x.$$

Therefore,

$$\frac{d}{dx}(\cot x) = -\csc^2 x.$$

Example 4.3-3: Given that $\frac{d}{dx}(\cos x) = -\sin x$, prove that $\frac{d}{dx}(\sec x) = \sec x \tan x$.
By definition,

$$\sec x = \frac{1}{\cos x}.$$

Thus,

$$\frac{d}{dx}(\sec x) = \frac{d}{dx}\left(\frac{1}{\cos x}\right).$$

Applying the quotient rule,

$$\frac{d}{dx}\left(\frac{1}{\cos x}\right) = \frac{\frac{d}{dx}(1)\cos x - \frac{d}{dx}(\cos x)}{\cos^2 x}.$$

Note that the derivative of a constant is 0. Let c be a constant, then we can write c as $c = cx^0$. Applying the power rule,

$$\frac{d}{dx}(c) = \frac{d}{dx}\left(cx^0\right) = c \times 0x^{0-1} = 0.$$

Also, since

$$\frac{d}{dx}(\cos x) = -\sin x,$$

we have

$$\frac{\frac{d}{dx}(1)\cos x - \frac{d}{dx}(\cos x)}{\cos^2 x} = \frac{\sin x}{\cos^2 x} = \sec x \tan x.$$

Therefore,

$$\frac{d}{dx}(\sec x) = \sec x \tan x.$$

Example 4.3-4: Given that $\frac{d}{dx}(\sin x) = \cos x$, prove that $\frac{d}{dx}(\csc x) = -\csc x \cot x$.
By definition,

$$\csc x = \frac{1}{\sin x}.$$

Thus,

$$\frac{d}{dx}(\csc x) = \frac{d}{dx}\left(\frac{1}{\sin x}\right).$$

Applying the quotient rule,

$$\frac{d}{dx}\left(\frac{1}{\sin x}\right) = \frac{\frac{d}{dx}(1)\sin x - \frac{d}{dx}(\sin x)}{\sin^2 x}.$$

Since

$$\frac{d}{dx}(1) = 0$$

and

$$\frac{d}{dx}(\sin x) = \cos x,$$

we have

$$\frac{\frac{d}{dx}(1)\sin x - \frac{d}{dx}(\sin x)}{\sin^2 x} = \frac{-\cos x}{\sin^2 x} = -\csc x \cot x.$$

Therefore,

$$\frac{d}{dx}(\csc x) = -\csc x \cot x.$$

Example 4.3-5: Given that $\frac{d}{dx}(\sinh x) = \cosh x$ and $\frac{d}{dx}(\cosh x) = \sinh x$, prove that $\frac{d}{dx}(\tanh x) = \operatorname{sech}^2 x$.
By definition,

$$\tanh x = \frac{\sinh x}{\cosh x}.$$

Thus,

$$\frac{d}{dx}(\tanh x) = \frac{d}{dx}\left(\frac{\sinh x}{\cosh x}\right).$$

Applying the quotient rule,

$$\frac{d}{dx}\left(\frac{\sinh x}{\cosh x}\right) = \frac{\frac{d}{dx}(\sinh x)\cosh x - \sinh x \frac{d}{dx}(\cosh x)}{\cosh^2 x}.$$

Since

$$\frac{d}{dx}(\sinh x) = \cosh x$$

and

$$\frac{d}{dx}(\cosh x) = \sinh x,$$

we have

$$\frac{\frac{d}{dx}(\sinh x)\cosh x - \sinh x \frac{d}{dx}(\cosh x)}{\cosh^2 x} = \frac{\cosh^2 x - \sinh^2 x}{\cosh^2 x}.$$

Since we have the identity $\cosh^2 x - \sinh^2 x = 1$,

$$\frac{\cosh^2 x - \sinh^2 x}{\cosh^2 x} = \frac{1}{\cosh^2 x} = \operatorname{sech}^2 x.$$

Therefore,

$$\frac{d}{dx}(\tanh x) = \operatorname{sech}^2 x.$$

Example 4.3-6: Given that $\frac{d}{dx}(\sinh x) = \cosh x$ and $\frac{d}{dx}(\cosh x) = \sinh x$, prove that $\frac{d}{dx}(\coth x) = -\operatorname{csch}^2 x$.
By definition,

$$\coth x = \frac{\cosh x}{\sinh x}.$$

Thus,

$$\frac{d}{dx}(\coth x) = \frac{d}{dx}\left(\frac{\cosh x}{\sinh x}\right).$$

Applying the quotient rule,

$$\frac{d}{dx}\left(\frac{\cosh x}{\sinh x}\right) = \frac{\frac{d}{dx}(\cosh x)\sinh x - \cosh x \frac{d}{dx}(\sinh x)}{\sinh^2 x}.$$

Since

$$\frac{d}{dx}(\sinh x) = \cosh x$$

and

$$\frac{d}{dx}(\cosh x) = \sinh x,$$

we have

$$\frac{\frac{d}{dx}(\cosh x)\sinh x - \cosh x \frac{d}{dx}(\sinh x)}{\sinh^2 x} = \frac{\sinh^2 x - \cosh^2 x}{\sinh^2 x}.$$

Since we have the identity $\cosh^2 x - \sinh^2 x = 1$,

$$\frac{\sinh^2 x - \cosh^2 x}{\sinh^2 x} = \frac{-1}{\sinh^2 x} = -\operatorname{csch}^2 x.$$

Therefore,

$$\frac{d}{dx}(\coth x) = -\operatorname{csch}^2 x.$$

Example 4.3-7: Given that $\frac{d}{dx}(\cosh x) = \sinh x$, prove that $\frac{d}{dx}(\operatorname{sech} x) = -\operatorname{sech} x \tanh x$.
By definition,

$$\operatorname{sech} x = \frac{1}{\cosh x}.$$

Thus,

$$\frac{d}{dx}(\operatorname{sech} x) = \frac{d}{dx}\left(\frac{1}{\cosh x}\right).$$

Applying the quotient rule,

$$\frac{d}{dx}\left(\frac{1}{\cosh x}\right) = \frac{\frac{d}{dx}(1)\cosh x - \frac{d}{dx}(\cosh x)}{\cosh^2 x}.$$

Since

$$\frac{d}{dx}(1) = 0$$

and

$$\frac{d}{dx}(\cosh x) = \sinh x,$$

we have

$$\frac{\frac{d}{dx}(1)\cosh x - \frac{d}{dx}(\cosh x)}{\cosh^2 x} = \frac{-\sinh x}{\cosh^2 x} = -\operatorname{sech} x \tanh x.$$

Therefore,

$$\frac{d}{dx}(\operatorname{sech} x) = -\operatorname{sech} x \tanh x.$$

Example 4.3-8: Given that $\frac{d}{dx}(\sinh x) = \cosh x$, prove that $\frac{d}{dx}(\operatorname{csch} x) = -\operatorname{csch} x \coth x$.
By definition,

$$\operatorname{csch} x = \frac{1}{\sinh x}.$$

Thus,

$$\frac{d}{dx}(\operatorname{csch} x) = \frac{d}{dx}\left(\frac{1}{\sinh x}\right).$$

Applying the quotient rule,

$$\frac{d}{dx}\left(\frac{1}{\sinh x}\right) = \frac{\frac{d}{dx}(1)\sinh x - \frac{d}{dx}(\sinh x)}{\sinh^2 x}.$$

Since

$$\frac{d}{dx}(1) = 0$$

and

$$\frac{d}{dx}(\sinh x) = \cosh x,$$

we have

$$\frac{\frac{d}{dx}(1)\sinh x - \frac{d}{dx}(\sinh x)}{\sinh^2 x} = \frac{-\cosh x}{\sinh^2 x} = -\operatorname{csch} x \coth x.$$

Therefore,

$$\frac{d}{dx}(\operatorname{csch} x) = -\operatorname{csch} x \coth x.$$

Example 4.3-9: Find the slope of the tangent line to the curve $y = \frac{\sin x}{x}$ at $x = 1$.
To find the slope of the tangent line to the curve, we need to find the

derivative of y with respect to x, $\frac{dy}{dx}$.
First, we have

$$\frac{dy}{dx} = \frac{d}{dx}\left(\frac{\sin x}{x}\right).$$

Applying the quotient rule,

$$\frac{d}{dx}\left(\frac{\sin x}{x}\right) = \frac{\frac{d}{dx}(\sin x)x - \sin x\frac{d}{dx}(x)}{x^2}$$

Since

$$\frac{d}{dx}(x) = 1$$

and

$$\frac{d}{dx}(\sin x) = \cos x,$$

we have

$$\frac{\frac{d}{dx}(\sin x)x - \sin x\frac{d}{dx}(x)}{x^2} = \frac{x\cos x - \sin x}{x^2}.$$

Plugging in $x = 1$, our final answer is

$$\cos 1 - \sin 1.$$

4.4 Chain Rule

In this section, we will discuss the chain rule. The chain rule is the rule for evaluating derivatives of compositions of functions. However, remember that the chain rule discussed in this section is only applied for ordinary derivatives of single variable functions.

Chain Rule
For two functions in terms of x, $u(x)$ and $v(x)$,

$$\frac{d}{dx}(v(u(x))) = \frac{d}{du}(v(u(x))) \times \frac{d}{dx}(u(x)).$$

The notation

$$\frac{d}{du}$$

means that we are taking derivatives with respect to u. So far, we have only used

$$\frac{d}{dx}$$

because we have only been evaluating derivatives with respect to x. The variable next to d in the denominator does not have to be x. It can be any variable, and we take derivative (or differentiate) with respect to whatever variable is next to the d in the denominator.

To apply the chain rule more easily, we will do u - substition (or v - substitution like in example 4.4-2 if we already used u). The u would represent the inner function of a composition of two functions. Now let us get into a few examples using the chain rule.

Example 4.4-1: Let $f(x) = \sin\left(x^2\right)$. Find $f'(3)$.
First, we have

$$f'(x) = \frac{d}{dx}\left(\sin\left(x^2\right)\right).$$

Let $u = x^2$,

$$\frac{d}{dx}\left(\sin\left(x^2\right)\right) = \frac{d}{dx}\left(\sin u\right).$$

Applying the chain rule,

$$\frac{d}{dx}\left(\sin u\right) = \frac{d}{du}(\sin u) \times \frac{du}{dx} = \cos u \times \frac{du}{dx}.$$

Since $u = x^2$,

$$\cos u \times \frac{du}{dx} = \cos\left(x^2\right) \times \frac{d}{dx}\left(x^2\right) = 2x\cos\left(x^2\right).$$

Plugging in $x = 3$, our final answer is

$$f'(3) = 6\cos(9).$$

Example 4.4-2: Evaluate $\frac{d}{dx}\left(e^{\sqrt{\cosh x}}\right)$.
Let $u = \sqrt{\cosh x}$,

$$\frac{d}{dx}\left(e^{\sqrt{\cosh x}}\right) = \frac{d}{dx}\left(e^u\right).$$

Applying the chain rule,

$$\frac{d}{dx}\left(e^u\right) = \frac{d}{du}\left(e^u\right) \times \frac{du}{dx} = e^u \times \frac{du}{dx}.$$

Since $u = \sqrt{\cosh x}$,

$$e^u \times \frac{du}{dx} = e^{\sqrt{\cosh x}} \times \frac{d}{dx}\left(\sqrt{\cosh x}\right).$$

Now let $v = \cosh x$,

$$\frac{d}{dx}\left(\sqrt{\cosh x}\right) = \frac{d}{dx}\left(\sqrt{v}\right).$$

Applying the chain rule,

$$\frac{d}{dx}\left(\sqrt{v}\right)=\frac{d}{dv}\left(\sqrt{v}\right)\times\frac{dv}{dx}=\frac{1}{2\sqrt{v}}\times\frac{dv}{dx}.$$

Since $v=\cosh x$,

$$\frac{1}{2\sqrt{v}}\times\frac{dv}{dx}=\frac{1}{2\sqrt{\cosh x}}\times\frac{d}{dx}(\cosh x)=\frac{\sinh x}{2\sqrt{\cosh x}}.$$

Therefore, our final answer is

$$\frac{d}{dx}\left(e^{\sqrt{\cosh x}}\right)=\frac{e^{\sqrt{\cosh x}}\sinh x}{2\sqrt{\cosh x}}.$$

Example 4.4-3: Find the slope of the tangent line to the curve $y=\tanh^{-1}(2x)$ at $x=\frac{1}{4}$.
First, we need to find the derivative of y with respect to x:

$$\frac{dy}{dx}=\frac{d}{dx}\left(\tanh^{-1}(2x)\right).$$

Let $u=2x$,

$$\frac{d}{dx}\left(\tanh^{-1}(2x)\right)=\frac{d}{dx}\left(\tanh^{-1}u\right).$$

Applying the chain rule,

$$\frac{d}{dx}\left(\tanh^{-1}u\right)=\frac{d}{du}\left(\tanh^{-1}u\right)\times\frac{du}{dx}=\frac{1}{1-u^2}\times\frac{du}{dx}.$$

Since $u=2x$,

$$\frac{1}{1-u^2}\times\frac{du}{dx}=\frac{1}{1-4x^2}\times\frac{d}{dx}(2x)=\frac{2}{1-4x^2}.$$

Plugging in $x=\frac{1}{4}$, our final answer is $\frac{8}{3}$.

Example 4.4-4: Evaluate $\frac{d}{dx}\left(\left|\sinh^{-1}x\right|\right)$.
Let $u=\sinh^{-1}x$,

$$\frac{d}{dx}\left(\left|\sinh^{-1}x\right|\right)=\frac{d}{dx}\left(|u|\right).$$

Applying the chain rule,

$$\frac{d}{dx}(|u|)=\frac{d}{du}(|u|)\times\frac{du}{dx}=\frac{u}{|u|}\times\frac{du}{dx}.$$

Since $u=\sinh^{-1}x$,

$$\frac{u}{|u|}\times\frac{du}{dx}=\frac{\sinh^{-1}x}{\left|\sinh^{-1}x\right|}\times\frac{d}{dx}\left(\sinh^{-1}x\right)=\frac{\sinh^{-1}x}{\left|\sinh^{-1}x\right|\sqrt{1+x^2}}.$$

Therefore, our final answer is

$$\frac{d}{dx}\left(\left|\sinh^{-1} x\right|\right) = \frac{\sinh^{-1} x}{\left|\sinh^{-1} x\right| \sqrt{1+x^2}}.$$

Example 4.4-5: Find the slope of the tangent line to the curve $y = |\ln x|$ at $x = 4$.
First, we need to find the derivative of y with respect to x:

$$\frac{dy}{dx} = \frac{d}{dx}\left(|\ln x|\right).$$

Let $u = \ln x$,

$$\frac{d}{dx}\left(|\ln x|\right) = \frac{d}{dx}\left(|u|\right).$$

Applying the chain rule,

$$\frac{d}{dx}\left(|u|\right) = \frac{d}{du}(|u|) \times \frac{du}{dx} = \frac{u}{|u|} \times \frac{du}{dx}.$$

Since $u = \ln x$,

$$\frac{u}{|u|} \times \frac{du}{dx} = \frac{\ln x}{|\ln x|} \times \frac{d}{dx}(\ln x) = \frac{\ln x}{x\,|\ln x|}.$$

Plugging in $x = 4$, the slope of the tangent line at $x = 4$ is

$$\frac{\ln 4}{4\,|\ln 4|}.$$

Since $\ln 4 > 0$, $|\ln 4| = \ln 4$. Therefore, our final answer is $\frac{1}{4}$.

Exercise Problems

1. Find the slope of the tangent line to the curve $y = x|x|$ at $x = -1$.

2. Evaluate $\frac{d}{dx}(\log_x 6)$.

3. Let $f(x) = \ln(|x|)$. Find $f'(x)$ and $f'(-1)$.

4. Let $f(x) = |x^3|$. Find $f'(x)$ and $f'(-2)$.

5. Define $F(x) = f(x)g(x)h(x)$ for all real x, and the functions $f(x)$, $g(x)$, and $h(x)$ are differentiable at $x = a$. Given that $F'(a) = 42F(a)$, $f'(a) = -14f(a)$, $g'(a) = 48g(a)$, and $h'(a) = \lambda h(a)$, find the value of λ.

6. Define $h(x) = \frac{f(x)}{g(x)}$ for all real x, and the functions $f(x)$ and $g(x)$ are differentiable. If $f'(x) = x^2 f(x)$, $g'(x) = xg(x)$, and $h'(b) = h(b)$, then what is/are the possible value(s) of b?

Chapter 5

Second Derivatives, Implicit Differentiation, and Logarithmic Differentiation

5.1 Second Derivatives

First of all, what is a second derivative? A second derivative is basically the derivative of a derivative. So far, we have only learned first derivatives, which are usually simply called "derivatives," because we have differentiated functions only once for each function. When we take a second derivative of a function, we differentiate that function twice.

The notation for second derivatives is

$$\frac{d^2}{dx^2},$$

which is simply

$$\frac{d}{dx}\left(\frac{d}{dx}\right).$$

Also, the second derivative of a function $f(x)$ with respect to x can be denoted as

$$f''(x).$$

Now let us look at a few examples of second derivatives.

Example 5.1-1: Evaluate $\frac{d^2}{dx^2}(\ln x)$.
First, we have

$$\frac{d}{dx}(\ln x) = \frac{1}{x}. \tag{5.1}$$

To find the second derivative, we differentiate (5.1) again:

$$\frac{d^2}{dx^2}(\ln x) = \frac{d}{dx}\left(\frac{1}{x}\right).$$

Applying the power rule,

$$\frac{d}{dx}\left(\frac{1}{x}\right) = -\frac{1}{x^2}.$$

Therefore,

$$\frac{d^2}{dx^2}(\ln x) = -\frac{1}{x^2}.$$

Example 5.1-2: Let $f(x) = xe^x$. Find $f''(2)$.
First, we have

$$f''(x) = \frac{d^2}{dx^2}(xe^x).$$

Let us find the first derivative first by applying the product rule:

$$\frac{d}{dx}(xe^x) = \frac{d}{dx}(x)e^x + x\frac{d}{dx}(e^x) = e^x + xe^x. \tag{5.2}$$

Differentiating (5.2) again,

$$\frac{d^2}{dx^2}(xe^x) = \frac{d}{dx}(e^x + xe^x) = \frac{d}{dx}(e^x) + \frac{d}{dx}(xe^x) = 2e^x + xe^x.$$

Plugging in $x = 2$, our final answer is

$$f''(2) = 4e^2.$$

5.2 Implicit Differentiation

So far, we have only learned how to differentiate y with respect to x for explicit equations, which are equations of the form

$$y = f(x)$$

where $f(x)$ is an expression in terms of x.

Implicit equations are equations that are not explicit. In this section, we will learn implicit differentiation, which is how to differentiate y with respect to x - or how to find $\frac{dy}{dx}$ - for implicit equations. Basically, we find $\frac{dy}{dx}$ for an implicit equation by differentiating both sides of the implicit equation with respect to x. Let us look at some examples.

Example 5.2-1: Find $\frac{dy}{dx}$ for $x = y^2x + y$.
Our equation is

$$x = y^2x + y \tag{5.3}$$

Differentiating both sides of (5.3) with respect to x,

$$\frac{d}{dx}(x) = \frac{d}{dx}(y^2x) + \frac{dy}{dx}. \tag{5.4}$$

Applying the power rule,

$$\frac{d}{dx}(x) = 1. \tag{5.5}$$

Applying the product rule,

$$\frac{d}{dx}(y^2x) = \frac{d}{dx}(y^2)x + y^2\frac{d}{dx}(x) = \frac{d}{dx}(y^2)x + y^2.$$

Applying the chain rule,

$$\frac{d}{dx}(y^2) = \frac{d}{dy}(y^2) \times \frac{dy}{dx} = 2y\frac{dy}{dx}.$$

Thus,

$$\frac{d}{dx}(y^2x) = 2xy\frac{dy}{dx} + y^2. \tag{5.6}$$

Plugging (5.6) and (5.5) into (5.4),

$$1 = 2xy\frac{dy}{dx} + y^2 + \frac{dy}{dx}.$$

Now we treat $\frac{dy}{dx}$ as a variable and solve for it:

$$1 - y^2 = (2xy + 1)\frac{dy}{dx}$$

$$\frac{1 - y^2}{2xy + 1} = \frac{dy}{dx}.$$

Therefore, our final answer is

$$\frac{dy}{dx} = \frac{1 - y^2}{2xy + 1}.$$

Example 5.2-2: Find $\frac{dy}{dx}$ for $\ln x = |y|e^y$.
Our equation is

$$\ln x = |y|e^y. \tag{5.7}$$

Differentiating both sides of (5.7) with respect to x,

$$\frac{d}{dx}(\ln x) = \frac{d}{dx}(|y|e^y). \tag{5.8}$$

We have

$$\frac{d}{dx}(\ln x) = \frac{1}{x}. \tag{5.9}$$

Applying the product rule,

$$\frac{d}{dx}(|y|e^y) = \frac{d}{dx}(|y|)\, e^y + |y|\frac{d}{dx}(e^y).$$

Applying the chain rule,

$$\frac{d}{dx}(|y|) = \frac{d}{dy}(|y|) \times \frac{dy}{dx} = \frac{y}{|y|}\frac{dy}{dx},$$

and

$$\frac{d}{dx}(e^y) = \frac{d}{dy}(e^y) \times \frac{dy}{dx} = e^y\frac{dy}{dx}.$$

Thus,

$$\frac{d}{dx}(|y|e^y) = \frac{ye^y}{|y|}\frac{dy}{dx} + |y|e^y\frac{dy}{dx} = \frac{ye^y}{|y|}\frac{dy}{dx} + \frac{|y|^2e^y}{|y|}\frac{dy}{dx}.$$

Since $|y|^2 = y^2$,

$$\frac{d}{dx}(|y|e^y) = \frac{ye^y}{|y|}\frac{dy}{dx} + \frac{y^2e^y}{|y|}\frac{dy}{dx}. \tag{5.10}$$

Plugging (5.9) and (5.10) into (5.8),

$$\frac{1}{x} = \frac{ye^y}{|y|}\frac{dy}{dx} + \frac{y^2e^y}{|y|}\frac{dy}{dx}.$$

Solving for $\frac{dy}{dx}$, our final answer is

$$\frac{dy}{dx} = \frac{|y|}{xye^y(1+y)}.$$

Example 5.2-3: Find the slope of the tangent line to the curve $\left|x^2+y\right| = x$ at the point $\left(\frac{1}{2}, \frac{-3}{4}\right)$.
To find the slope of the tangent line to the curve, we first need to find the derivative of y with respect to x, $\frac{dy}{dx}$.
Our equation is

$$\left|x^2+y\right| = x. \tag{5.11}$$

Differentiating both sides of (5.11) with respect to x,

$$\frac{d}{dx}\left(\left|x^2+y\right|\right) = \frac{d}{dx}(x). \tag{5.12}$$

Applying the power rule,

$$\frac{d}{dx}(x) = 1. \tag{5.13}$$

Let $u = x^2 + y$,

$$\frac{d}{dx}\left(\left|x^2+y\right|\right) = \frac{d}{dx}(|u|).$$

Applying the chain rule,

$$\frac{d}{dx}(|u|) = \frac{d}{du}(|u|) \times \frac{du}{dx} = \frac{u}{|u|}\frac{du}{dx}.$$

Since $u = x^2 + y$,

$$\frac{u}{|u|}\frac{du}{dx} = \frac{x^2+y}{|x^2+y|}\frac{d}{dx}\left(x^2+y\right).$$

Note that

$$\frac{d}{dx}\left(x^2+y\right) = \frac{d}{dx}\left(x^2\right) + \frac{dy}{dx} = 2x + \frac{dy}{dx}.$$

Thus,

$$\frac{d}{dx}\left(\left|x^2+y\right|\right) = \frac{2x\left(x^2+y\right)}{|x^2+y|} + \frac{x^2+y}{|x^2+y|}\frac{dy}{dx}. \tag{5.14}$$

Plugging (5.13) and (5.14) into (5.12),

$$\frac{2x\left(x^2+y\right)}{|x^2+y|} + \frac{x^2+y}{|x^2+y|}\frac{dy}{dx} = 1.$$

Solving for $\frac{dy}{dx}$,

$$\frac{dy}{dx} = \frac{\left|x^2 + y\right|}{x^2 + y} - 2x.$$

Plugging in $x = \frac{1}{2}$ and $y = \frac{-3}{4}$, the slope of the tangent line to the curve $\left|x^2 + y\right| = x$ at the point $\left(\frac{1}{2}, \frac{-3}{4}\right)$ is

$$\frac{\left|\left(\frac{1}{2}\right)^2 - \frac{3}{4}\right|}{\left(\frac{1}{2}\right)^2 - \frac{3}{4}} - 2 \times \frac{1}{2} = -2.$$

Example 5.2-4: Find the slope of the tangent line to the curve $\left(x^2 + y^2 - 1\right)^3 = x^2 y^3$ at the point $(1, 1)$.
Our equation is

$$\left(x^2 + y^2 - 1\right)^3 = x^2 y^3. \tag{5.15}$$

Differentiating both sides of (5.15) with respect to x,

$$\frac{d}{dx}\left(\left(x^2 + y^2 - 1\right)^3\right) = \frac{d}{dx}\left(x^2 y^3\right). \tag{5.16}$$

Let $u = x^2 + y^2 - 1$,

$$\frac{d}{dx}\left(\left(x^2 + y^2 - 1\right)^3\right) = \frac{d}{dx}\left(u^3\right).$$

Applying the chain rule,

$$\frac{d}{dx}\left(u^3\right) = \frac{d}{du}\left(u^3\right) \times \frac{du}{dx} = 3u^2\frac{du}{dx}.$$

Since $u = x^2 + y^2 - 1$,

$$3u^2\frac{du}{dx} = 3\left(x^2 + y^2 - 1\right)^2 \times \frac{d}{dx}\left(x^2 + y^2 - 1\right).$$

Note that

$$\frac{d}{dx}\left(x^2 + y^2 - 1\right) = \frac{d}{dx}\left(x^2\right) + \frac{d}{dx}\left(y^2\right) - \frac{d}{dx}(1) = 2x + \frac{d}{dx}\left(y^2\right).$$

Applying the chain rule,

$$\frac{d}{dx}\left(y^2\right) = \frac{d}{dy}\left(y^2\right) \times \frac{dy}{dx} = 2y\frac{dy}{dx}.$$

So, we have

$$\frac{d}{dx}\left(x^2 + y^2 - 1\right) = 2x + 2y\frac{dy}{dx}.$$

Thus,

$$\frac{d}{dx}\left(\left(x^2+y^2-1\right)^3\right)=6x\left(x^2+y^2-1\right)^2+6y\left(x^2+y^2-1\right)^2\frac{dy}{dx}. \quad (5.17)$$

Applying the product rule,

$$\frac{d}{dx}\left(x^2y^3\right)=\frac{d}{dx}\left(x^2\right)y^3+x^2\frac{d}{dx}\left(y^3\right)=2xy^3+x^2\frac{d}{dx}\left(y^3\right).$$

Applying the chain rule,

$$\frac{d}{dx}\left(y^3\right)=\frac{d}{dy}\left(y^3\right)\times\frac{dy}{dx}=3y^2\frac{dy}{dx}.$$

Thus,

$$\frac{d}{dx}\left(x^2y^3\right)=2xy^3+3x^2y^2\frac{dy}{dx}. \quad (5.18)$$

Plugging (5.17) and (5.18) into (5.16),

$$6x\left(x^2+y^2-1\right)^2+6y\left(x^2+y^2-1\right)^2\frac{dy}{dx}=2xy^3+3x^2y^2\frac{dy}{dx}.$$

Solving for $\frac{dy}{dx}$,

$$\frac{dy}{dx}=\frac{6x\left(x^2+y^2-1\right)^2-2xy^3}{3x^2y^2-6y\left(x^2+y^2-1\right)^2}.$$

Plugging in $x=1$ and $y=1$, the slope of the tangent line to the curve $\left(x^2+y^2-1\right)^3=x^2y^3$ at the point $(1,1)$ is

$$\frac{6-2}{3-6}=-\frac{4}{3}.$$

By the way, $\left(x^2+y^2-1\right)^3=x^2y^3$ is a heart curve.

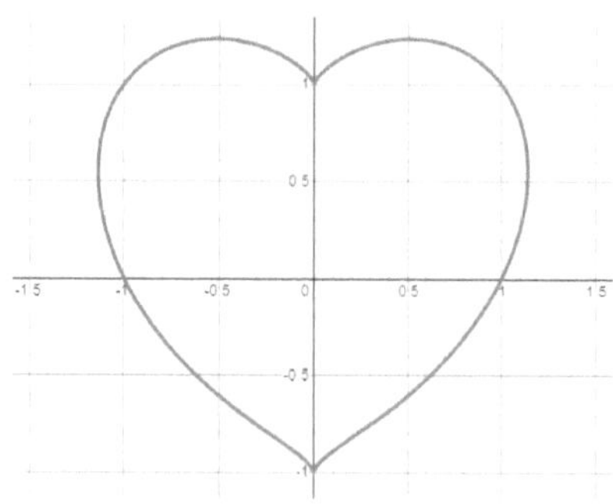

Figure 5.1: Graph of $\left(x^2+y^2-1\right)^3=x^2y^3$

5.3 Logarithmic Differentiation

In this section, we will discuss a method of differentiation called logarithmic differentiation. Logarithmic differentiation is a method of differentiation in which we take natural logarithm on both sides of an equation and then perform implicit differentiation, which we learned in the last section.

Now let us get into a few examples of problems using logarithmic differentiation.

Example 5.3-1: Evaluate $\frac{d}{dx}\left(x^x\right)$.
First, let

$$y = x^x. \tag{5.19}$$

Then, to find

$$\frac{d}{dx}\left(x^x\right),$$

we need to find $\frac{dy}{dx}$. Taking natural logarithm of both sides of (5.19),

$$\ln y = x \ln x. \tag{5.20}$$

Differentiating both sides of (5.20) with respect to x,

$$\frac{d}{dx}\left(\ln y\right) = \frac{d}{dx}\left(x \ln x\right). \tag{5.21}$$

Applying the chain rule,

$$\frac{d}{dx}\left(\ln y\right) = \frac{d}{dy}\left(\ln y\right) \times \frac{dy}{dx} = \frac{1}{y}\frac{dy}{dx}. \tag{5.22}$$

Applying the product rule,

$$\frac{d}{dx}\left(x \ln x\right) = \frac{d}{dx}(x) \ln x + x\frac{d}{dx}\left(\ln x\right) = \ln x + 1. \tag{5.23}$$

Plugging (5.22) and (5.23) into (5.21),

$$\frac{1}{y}\frac{dy}{dx} = \ln x + 1.$$

Solving for $\frac{dy}{dx}$,

$$\frac{dy}{dx} = y\left(\ln x + 1\right).$$

Since we have $y = x^x$, our final answer is

$$\frac{d}{dx}\left(x^x\right) = x^x\left(\ln x + 1\right).$$

Example 5.3-2: Find the slope of the tangent line to the curve $y = \sqrt[x]{x}$ at the point $x = \frac{1}{e}$.
To find the slope of the tangent line, we need to find $\frac{dy}{dx}$. First, note that

$$\sqrt[x]{x} = x^{\frac{1}{x}}.$$

Taking natural logarithm of both sides of

$$y = x^{\frac{1}{x}},$$

we have

$$\ln y = \frac{\ln x}{x}. \tag{5.24}$$

Differentiating both sides of (5.24) with respect to x,

$$\frac{d}{dx}(\ln y) = \frac{d}{dx}\left(\frac{\ln x}{x}\right). \tag{5.25}$$

Applying the chain rule,

$$\frac{d}{dx}(\ln y) = \frac{d}{dy}(\ln y) \times \frac{dy}{dx} = \frac{1}{y}\frac{dy}{dx}. \tag{5.26}$$

Applying the quotient rule,

$$\frac{d}{dx}\left(\frac{\ln x}{x}\right) = \frac{\frac{d}{dx}(\ln x)\,x - \ln x \frac{d}{dx}(x)}{x^2} = \frac{1-\ln x}{x^2}. \tag{5.27}$$

Plugging (5.26) and (5.27) into (5.25),

$$\frac{1}{y}\frac{dy}{dx} = \frac{1-\ln x}{x^2}.$$

Solving for $\frac{dy}{dx}$,

$$\frac{dy}{dx} = y\left(\frac{1-\ln x}{x^2}\right).$$

Since $y = x^{\frac{1}{x}}$,

$$\frac{dy}{dx} = x^{\frac{1}{x}}\left(\frac{1-\ln x}{x^2}\right).$$

Plugging in $x = \frac{1}{e}$, the slope of the tangent line to the curve $y = \sqrt[x]{x}$ at the point $x = \frac{1}{e}$ is

$$\frac{1}{e^e} \times \frac{2}{e^{-2}} = 2e^{2-e}.$$

Exercise Problems

1. Given that $x^y = y^x$, find $\frac{dy}{dx}$.

2. If $\frac{d}{dx}(g(x)) = h(x)$ and $\frac{d}{dx}(h(x)) = g\left(x^2\right)$, evaluate $\frac{d^2}{dx^2}\left(g\left(x^3\right)\right)$.

3. Given $y = f(x)$, $\eta = f'(x)$, and $\tau = f''(x)$, find $\frac{d^2x}{dy^2}$ in terms of η and τ.

4. Find the slope of the tangent line to the curve $x^2 + y^2 = 1$ at the point (0,1).

5. Prove that
$$\frac{d}{dx}(\arcsin x) = \frac{1}{\sqrt{1-x^2}}$$
using implicit differentiation. (Hint: use $\sin(\arcsin x) = x$)

6. Prove that
$$\frac{d}{dx}\left(\sinh^{-1} x\right) = \frac{1}{\sqrt{1+x^2}}$$
using implicit differentiation. (Hint: use $\sinh(\sinh^{-1} x) = x$)

7.

Lambert-W Function
By definition, Lambert-W function, $W(x)$, is the inverse of xe^x. Its domain is $\left[-\frac{1}{e}, \infty\right)$, and its range is $[-1, \infty)$.

Evaluate $\frac{d}{dx}(W(x))$.

Chapter 6

Analyzing Single Variable Functions

6.1 Introduction

In this chapter, we will discuss some important features of a single variable function, which is a function in terms of only one variable, e.g. $f(x) = x^2$. Such features are the intervals where a function is increasing or decreasing, critical points of a function, and inflection points of a function. We will learn how to apply the concepts of ordinary derivatives we have learned so far to analyze those features.

Those features are also present in multivariable functions. However, to analyze those features for a multivariable function, we need to use partial derivatives. Although partial derivatives will be briefly introduced in chapter 14 to give you a taste of what you can expect ahead if you continue learning calculus, application of partial derivatives in analyzing multivariable functions will not be discussed in this book. You can, however, learn it by yourself after reading this book if you want to.

Also, there is one final note I need to make before we begin to discuss the concepts of this chapter. In this chapter, I will not give you any formal definitions. I will only explain things intuitively. I made this decision based on my experience when I first learned calculus. What helped me the most in learning these concepts are not formal definitions, but rather building an intuition for them and getting used to them through examples.

6.2 Increasing and Decreasing at a Point

First, we will discuss how to determine if a function is increasing or decreasing at a certain point.

Recall that the derivative of a function at a certain point is the slope of the tangent line to the curve of the function at that point. If the slope of the tangent line to the curve of a function at a certain point is positive, then that means the function is increasing at that point. If the slope of the tangent line to the curve of a function at a certain point is negative, then that means the function is decreasing at that point.

Therefore, to determine if a function is increasing or decreasing, we simply determine if the derivative of the function is positive or negative. Now let us look at some examples.

Example 6.2-1: Determine if the graph of $y = x^2$ is increasing or decreasing at $x = 2$.
I know that this can be done without using calculus because it is simply a parabola, but let us start with an easy one to see how it can be done

using calculus.
First, the derivative is

$$y' = 2x$$

by applying the power rule. Plugging in $x = 2$, $y'(2) = 4$. Since the derivative is positive, the function is increasing at $x = 2$.

Example 6.2-2: Determine if the graph of $y = \ln x$ is increasing or decreasing at $x = \frac{1}{e}$.
The derivative is

$$y' = \frac{1}{x}.$$

Plugging in $x = e^{-1}$, $y'\left(e^{-1}\right) = e$. Since the derivative is positive, the function is increasing at $x = e^{-1}$.

Example 6.2-3: Determine if the graph of $y = x^4 - \frac{5}{2}x^2 + 1$ is increasing or decreasing at $x = \frac{1}{4}$.
Applying the power rule, the derivative is

$$y' = 4x^3 - 5x.$$

Plugging in $x = \frac{1}{4}$,

$$y'\left(\frac{1}{4}\right) = \frac{1}{16} - \frac{5}{4} = \frac{-19}{16}.$$

Since the derivative is negative, the function is decreasing at $x = \frac{1}{4}$.

Example 6.2-4: Determine if the graph of $y = \sqrt[x]{x}$ is increasing or decreasing at $x = 5$.
In example 5.3-2, we found that the derivative of this curve is

$$y' = \sqrt[x]{x}\left(\frac{1 - \ln x}{x^2}\right).$$

Plugging in $x = 5$,

$$y'(5) = \sqrt[5]{5}\left(\frac{1 - \ln 5}{25}\right).$$

Because $1 - \ln 5 < 0$,

$$\sqrt[5]{5}\left(\frac{1 - \ln 5}{25}\right) < 0.$$

Since the derivative is negative, the function is decreasing at $x = 5$.

6.3 Critical Points

In this section, we will discuss how to find critical points for a function. Critical points of a function are the points where extrema (plural of *extremum*), i.e. maxima (plural of *maximum*) and minima (plural of *minimum*), of the function might occur. We will discuss how to find extrema later in the next section.

> **Critical Points**
> Critical points of a function are where the derivative of the function is zero or undefined and where the function is defined.

Example 6.3-1: Find the critical point(s) of $f(x) = x^2$.
By applying the power rule, the derivative of the function is

$$f'(x) = 2x.$$

Solving for the point(s) where the derivative is 0,

$$\begin{aligned} f'(x) &= 0 \\ 2x &= 0 \\ x &= 0 \end{aligned}$$

is where there might be a critical point of the function. There is no point where the derivative is undefined. Also, the function $f(x) = x^2$ is defined at $x = 0$. Therefore, the critical point of $f(x) = x^2$ is at $x = 0$.

Example 6.3-2: Find the critical point(s) of $f(x) = \sqrt{x}$.
By applying the power rule, the derivative of the function is

$$f'(x) = \frac{1}{2\sqrt{x}}.$$

There is no point where the derivative of the function is zero. The derivative is undefined at $x = 0$. Also, the function $f(x) = \sqrt{x}$ is defined at $x = 0$. Therefore, the critical point of $f(x) = \sqrt{x}$ is at $x = 0$.

Example 6.3-3: Does the function $f(x) = \ln x$ have critical points?
The derivative of the function is

$$f'(x) = \frac{1}{x}.$$

There is no point where the derivative is zero. The derivative is undefined at $x = 0$. However, the function $f(x) = \ln x$ is undefined at $x = 0$. Therefore, the function $f(x) = \ln x$ does not have any critical points.

Example 6.3-4: Find the critical point(s) of $f(x) = x^x$.
The domain of the function

$$f(x) = x^x = e^{\ln(x^x)} = e^{x \ln x}$$

is $(0, \infty)$ because $\ln x$ is defined for $x \in (0, \infty)$.
In example 5.3-1, we found that the derivative of the function is

$$f'(x) = x^x(\ln x + 1).$$

Now we need to find the point(s) where the derivative is zero:

$$x^x(\ln x + 1) = 0. \tag{6.1}$$

Since $x^x \neq 0$ for $x \in (0, \infty)$, we can divide both sides of (6.1) by x^x:

$$\begin{aligned} \ln x + 1 &= 0 \\ \ln x &= -1 \\ x &= e^{-1}. \end{aligned}$$

So, $x = e^{-1}$ is where there might be a critical point of $f(x) = x^x$.
The derivative

$$f'(x) = x^x(\ln x + 1)$$

is undefined at $x = 0$ because x^x is undefined at $x = 0$.
However, the function $f(x) = x^x$ is undefined at $x = 0$, so $x = 0$ is not a critical point of $f(x) = x^x$. Therefore, the critical point of $f(x) = x^x$ is at $x = e^{-1}$.

6.4 Maxima and Minima of Functions

In this section, we will learn how to find extrema, i.e. maxima and minima, of a function. Before you start to learn calculus, you should have already learned the skill to plot functions on a graphing calculator to find maxima and minima, so I will not explain what maxima and minima are.

To find the maxima and minima, we need to find critical points first, which is what we just learned in the last section. If the function increases on the interval to the left of the critical point and decreases on the interval to the right of the critical point, then that critical point is a maximum. If the function decreases on the interval to the left of the critical point and increases on the interval to the right of the critical point, then that critical point is a minimum.

This should be somewhat straightforward if you think of the parabola.

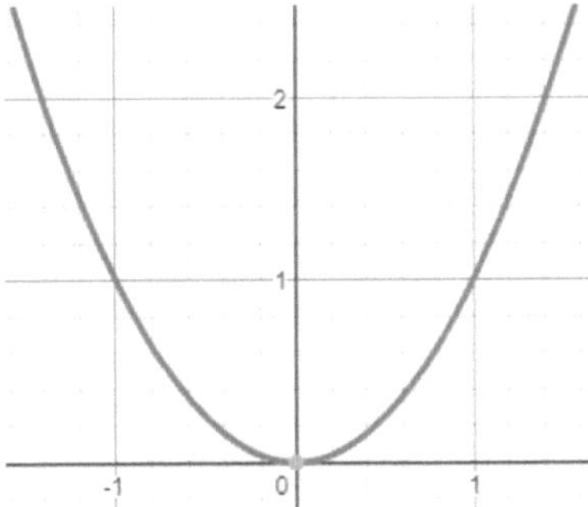

Figure 6.1: Graph of $f(x) = x^2$

From example 6.3-1, we know that the critical point of $f(x) = x^2$ is $x = 0$. We also know that the point at $x = 0$ is the minimum of $f(x) = x^2$ because it is the vertex of the parabola. You can notice that the function decreases on the interval to the left of the point at $x = 0$ and increases on the interval to the right of the point at $x = 0$. If you want to visualize the maximum, you can think of the graph of $y = -x^2$.

Also, from section 6.2, we learned how to determine if a function is increasing or decreasing by determining if the derivative is positive or negative. Now let us look at some examples using this method, and then I will introduce you to another method of finding extrema.

Example 6.4-1: Find the maxima and minima of the function $f(x) = x^4 - \frac{5}{2}x^2 + 1$.
First, let us find the critical points of the function. The derivative of the function is

$$f'(x) = 4x^3 - 5x.$$

Now let us find the points where the derivative is zero:

$$4x^3 - 5x = 0$$
$$x\left(4x^2 - 5\right) = 0.$$

So, the points where the derivative is zero are $x = -\frac{\sqrt{5}}{2}$, $x = 0$, and $x = \frac{\sqrt{5}}{2}$. There is no point where the derivative is undefined. Also, the function $f(x) = x^4 - \frac{5}{2}x^2 + 1$ is defined at $x = -\frac{\sqrt{5}}{2}$, $x = 0$, and $x = \frac{\sqrt{5}}{2}$. Therefore, the critical points are at $x = -\frac{\sqrt{5}}{2}$, $x = 0$, and $x = \frac{\sqrt{5}}{2}$.

Now let us determine if these critical points are maxima, minima, or neither.

For the critical point $x = -\frac{\sqrt{5}}{2}$, we need to determine if the derivative of the function is positive or negative on the intervals to the left and to the right of that point. To determine whether the derivative is positive or negative on the interval to the left of that point, we need to plug any x value less than $-\frac{\sqrt{5}}{2}$ into the derivative. We can plug in $x = -2$. Then, we will have

$$f'(-2) = 4(-2)^3 - 5(-2) = -22.$$

To determine whether the derivative is positive or negative on the interval to the right of that point, we need to plug any x value larger than $-\frac{\sqrt{5}}{2}$ and less than 0 into the derivative because 0 is the next critical point. We can plug in $x = -1$. Then, we will have

$$f'(-1) = 4(-1)^3 - 5(-1) = 1.$$

So, the function decreases to the left and increases to the right of the critical point $x = -\frac{\sqrt{5}}{2}$. That means the function has a minimum at $x = -\frac{\sqrt{5}}{2}$.

Now it is the turn for the critical point $x = 0$. We have already determined that the function increases on the interval to the left of $x = 0$ when we determined that the function increases on the interval to the right of $x = -\frac{\sqrt{5}}{2}$. To determine whether the derivative is positive or negative on the interval to the right of $x = 0$, we need to plug any x value larger than 0 and less than $\frac{\sqrt{5}}{2}$ into the derivative because $\frac{\sqrt{5}}{2}$ is the next critical point. We can plug in $x = 1$. Then, we will have

$$f'(1) = 4(1)^3 - 5(1) = -1.$$

So, the function increases to the left and decreases to the right of the critical point $x = 0$. That means the function has a maximum at $x = 0$.

Finally, we need to determine whether the critical point $x = \frac{\sqrt{5}}{2}$ is a maximum, a minimum, or neither. We have already determined that the function decreases on the interval to the left of $x = \frac{\sqrt{5}}{2}$ when we determined that the function decreases on the interval to the right of $x = 0$. To determine whether the derivative is positive or negative on the interval to the right of $x = \frac{\sqrt{5}}{2}$, we need to plug any x value larger than $\frac{\sqrt{5}}{2}$ into the derivative. We can plug in $x = 2$. Then, we will have

$$f'(2) = 4(2)^3 - 5(2) = 22.$$

So, the function decreases to the left and increases to the right of the critical point $x = \frac{\sqrt{5}}{2}$. That means the function has a minimum at $x = \frac{\sqrt{5}}{2}$.

Therefore, the function $f(x) = x^4 - \frac{5}{2}x^2 + 1$ has two minima at $x = -\frac{\sqrt{5}}{2}$

and $x = \frac{\sqrt{5}}{2}$ and one maximum at $x = 0$. I know that was a lot of words, but here is the graph for you to visualize better:

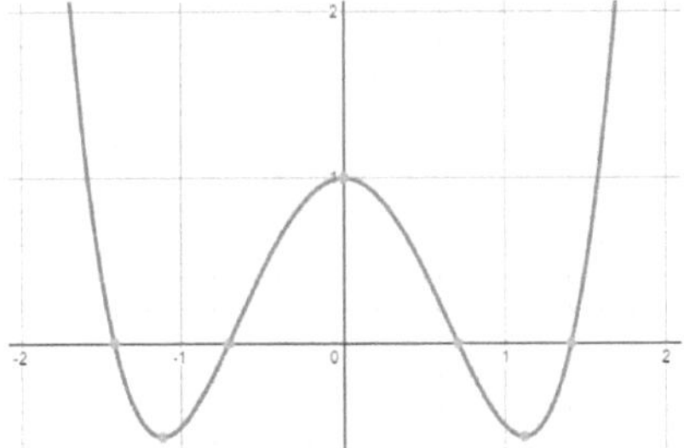

Figure 6.2: Graph of $f(x) = x^4 - \frac{5}{2}x^2 + 1$

Example 6.4-2: Find the extremum of the function $f(x) = x^x$.
From example 5.3-1, we found that the derivative is

$$f'(x) = x^x\left(\ln x + 1\right).$$

Also, from example 6.3-4, we found that the critical point of $f(x) = x^x$ is at $x = e^{-1}$.

Now we need to determine if the critical point is a maximum, a minimum, or neither. To determine if the derivative is positive or negative on the interval to the left of $x = e^{-1}$, we need to plug any x value less than e^{-1} and larger than 0 into the derivative because the function is only defined for $x > 0$. We can plug in $x = \frac{1}{3}$. Then, we have

$$f'\left(\frac{1}{3}\right) = \frac{1}{\sqrt[3]{3}}\left(\ln\left(\frac{1}{3}\right) + 1\right).$$

Since $\frac{1}{3} < e^{-1}$, we have $\ln\left(\frac{1}{3}\right) < \ln\left(e^{-1}\right) = -1$. So,

$$\ln\left(\frac{1}{3}\right) + 1 < 0.$$

Also, $\frac{1}{\sqrt[3]{3}} > 0$, so we have

$$f'\left(\frac{1}{3}\right) = \frac{1}{\sqrt[3]{3}}\left(\ln\left(\frac{1}{3}\right) + 1\right) < 0.$$

To determine if the derivative is positive or negative on the interval to the right of $x = e^{-1}$, we need to plug any x value larger than e^{-1} into the derivative. We can plug in $x = 1$. Then, we have

$$f'(1) = 1^1(\ln 1 + 1) = 1 > 0.$$

So, the function decreases to the left and increases to the right of $x = e^{-1}$. That means the function has a minimum at the critical point at $x = e^{-1}$. Here is the graph of the function for you to visualize better:

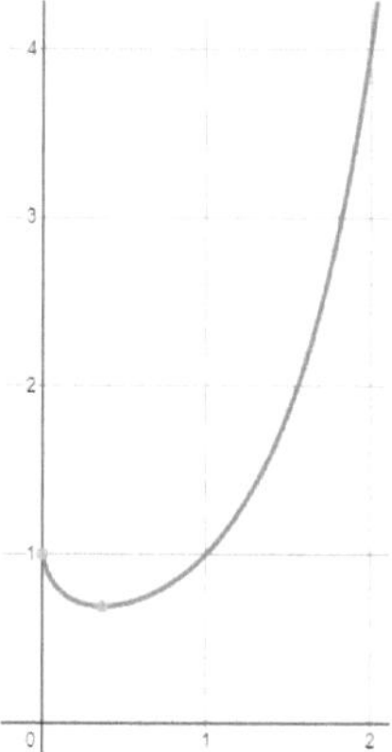

Figure 6.3: Graph of $f(x) = x^x$

Example 6.4-3: Determine if the function $f(x) = x^3$ has any maxima or minima.
The derivative of the function is

$$f'(x) = 3x^2.$$

So, the critical point is at $x = 0$. Now let us determine if this critical point is a maximum, a minimum, or neither.

To determine if the derivative is positive or negative on the interval to the left of $x = 0$, we need to plug any x value less than 0 into the derivative. We can plug in $x = -1$. Then, we have

$$f'(-1) = 3(-1)^2 = 3 > 0.$$

To determine if the derivative is positive or negative on the interval to the right of $x = 0$, we need to plug any x value larger than 0 into the derivative. We can plug in $x = 1$. Then, we have

$$f'(1) = 3(1)^2 = 3 > 0.$$

So, the function increases to the left and increases to the right of the critical point $x = 0$. That means the function has neither a maximum nor a minimum at the critical point at $x = 0$, so the function $f(x) = x^3$ does not have any maxima or minima.

Another way to find the extrema is to use the second derivative. As said before, a function increases to the left and decreases to the right of a maximum. So, going from left to right, the derivative is positive, then zero (or undefined) at the maximum (because it is a critical point), and then becomes negative. So, we can see that the derivative is decreasing at the maximum. Also, a function decreases to the left and increases to the right of a minimum. So, going from left to right, the derivative is negative, then zero (or undefined) at the minimum (because it is a critical point), and then becomes positive. So, we can see that the derivative is increasing at the minimum.

To determine if the function increases or decreases, we determine if the derivative is positive or negative. Similarly, to determine if the derivative increases or decreases, we determine if the derivative of the derivative, or the second derivative, is positive or negative. So, if the second derivative at a critical point is positive, then the derivative is increasing, and the critical point is a minimum. If the second derivative at a critical point is negative, then the derivative is decreasing, and the critical point is a maximum. Let us get back to the example 6.4-1 but with the method of using the second derivative.

Example 6.4-4: Find the extrema of the function $f(x) = x^4 - \frac{5}{2}x^2 + 1$.
From example 6.4-1, we found that the critical points are at $x = -\frac{\sqrt{5}}{2}$, $x = 0$, and $x = \frac{\sqrt{5}}{2}$.
The derivative of the function is

$$f'(x) = 4x^3 - 5x.$$

The second derivative is

$$f''(x) = 12x^2 - 5.$$

The second derivative at the critical point $x = -\frac{\sqrt{5}}{2}$ is

$$f''\left(-\frac{\sqrt{5}}{2}\right) = 12\left(-\frac{\sqrt{5}}{2}\right)^2 - 5 = 10 > 0,$$

so the function has a minimum at $x = -\frac{\sqrt{5}}{2}$.
The second derivative at the critical point $x = 0$ is

$$f''(0) = 12(0)^2 - 5 = -5 < 0,$$

so the function has a maximum at $x = 0$.
The second derivative at the critical point $x = \frac{\sqrt{5}}{2}$ is

$$f''\left(\frac{\sqrt{5}}{2}\right) = 12\left(\frac{\sqrt{5}}{2}\right)^2 - 5 = 10 > 0,$$

so the function has a minimum at $x = \frac{\sqrt{5}}{2}$.

6.5 Concave Up, Concave Down, and Inflection Points

In this section, we will discuss how to determine if a function is concave up or concave down on some interval. We will also discuss the inflection points.

> **Concave Up and Concave Down**
> • If the second derivative of a function is positive on an interval, then the function is concave up on that interval.
> • If the second derivative of a function is negative on an interval, then the function is concave down on that interval.

With this definition, it would be hard to visualize how concave up and concave down would look like. To visualize the shape of a concave up interval, you can think of the graph of $y = x^2$. The second derivative of $y = x^2$ is $y'' = 2$, which is always positive, so the parabola $y = x^2$ is concave up everywhere on its domain. Concave up intervals usually look like the shape of the parabola $y = x^2$ or part of the parabola $y = x^2$. To visualize the shape of a concave down interval, you can think of the graph of $y = -x^2$. The second derivative of $y = -x^2$ is $y'' = -2$, which is always negative, so the parabola $y = -x^2$ is concave down everywhere on its domain. Concave down intervals usually look like the shape of the parabola $y = -x^2$ or part of the parabola $y = -x^2$.

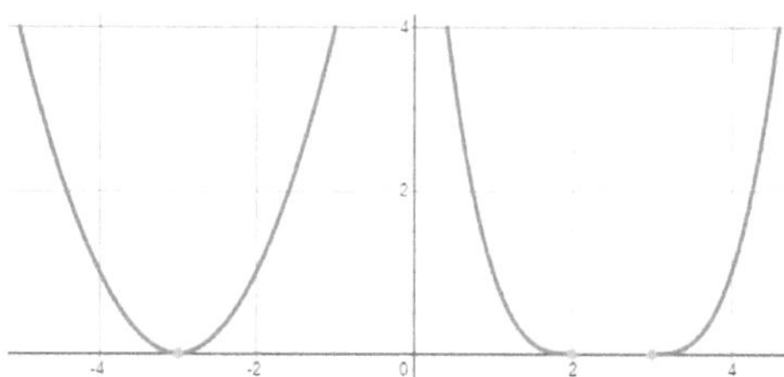

Figure 6.4: How concave up intervals might look

In general, concave up intervals are where the second derivative is positive, so the derivative (or the slope of the tangent line to the curve) increases. If you draw tangent lines to each graph in figure 6.4, you will see that the slope of the tangent lines increases as we go from left to right of each graph.

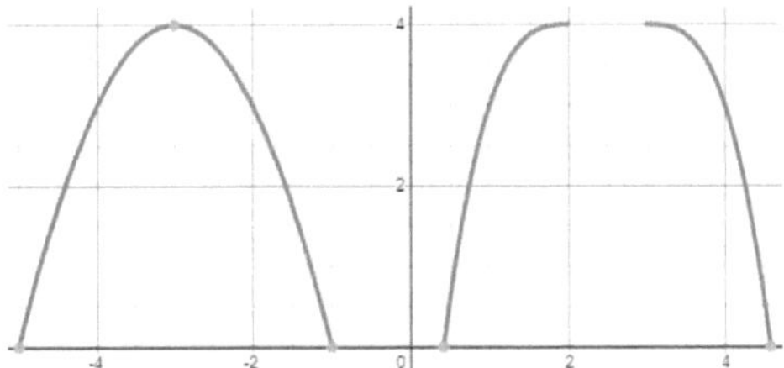

Figure 6.5: How concave down intervals might look

In general, concave down intervals are where the second derivative is negative, so the derivative (or the slope of the tangent line to the curve) decreases. If you draw tangent lines to each graph in figure 6.5, you will see that the slope of the tangent lines decreases as we go from left to right of each graph.

Here is a fun way to remember the general shapes of concave up and concave down. When the function is cheered up (concave up), it smiles (like in the left graph of figure 6.4). When the function is sad and down (concave down), it has a sad face, and its mouth goes down (like in the left graph of figure 6.5).

> **Inflection Points**
> Inflection points of a function are where the function changes from concave up to concave down or concave down to concave up.

Using the definition of concave up and concave down, we can also say that inflection points of a function are where the second derivative of the function changes from being positive to being negative or being negative to being positive. To find the inflection points, we need to find the points where the second derivative is zero.

Example 6.5-1: Find the inflection point(s), concave up interval(s), and concave down interval(s) of the function $f(x) = x^3$.
The derivative of the function is

$$f'(x) = 3x^2,$$

and the second derivative is

$$f''(x) = 6x.$$

So, the point where the second derivative is zero is $x = 0$. Now we need to determine if the function changes from concave up to concave down or concave down to concave up at that point.

To determine if the function is concave up or down on the interval to the left of $x = 0$, or to determine if the second derivative of the function is positive or negative on the interval to the left of $x = 0$, we need to plug any x value less than 0 into the second derivative. We can plug in $x = -1$. Then, we have

$$f''(-1) = 6(-1) = -6 < 0,$$

so the function is concave down to the left of $x = 0$.

To determine if the function is concave up or down on the interval to the right of $x = 0$, or to determine if the second derivative of the function is positive or negative on the interval to the right of $x = 0$, we need to plug any x value larger than 0 into the second derivative. We can plug in $x = 1$. Then, we have

$$f''(-1) = 6(1) = 6 > 0,$$

so the function is concave up to the right of $x = 0$.

Finally, the function $f(x) = x^3$ has an inflection point at $x = 0$ where the function changes from concave down to concave up. The concave down interval is $(-\infty, 0)$, and the concave up interval is $(0, \infty)$. If you want to visualize this better, you can draw the graph of $f(x) = x^3$, look at it, and compare parts of it with the graphs in figure 6.4 and figure 6.5.

Example 6.5-2: Find the inflection point(s), concave up interval(s), and concave down interval(s) of the function $f(x) = x^4 - \frac{5}{2}x^2 + 1$.
The derivative of the function is

$$f'(x) = 4x^3 - 5x,$$

and the second derivative is

$$f''(x) = 12x^2 - 5.$$

Solving for the points where the second derivative is zero, the inflection points are at $x = -\frac{\sqrt{15}}{6}$ and $x = \frac{\sqrt{15}}{6}$. Now let us determine if the intervals to the left and to the right of those inflection points are concave up or concave down.

To determine if the function is concave up or down on the interval to the left of $x = -\frac{\sqrt{15}}{6}$, or to determine if the second derivative of the function is positive or negative on the interval to the left of $x = -\frac{\sqrt{15}}{6}$, we need to plug any x value less than $-\frac{\sqrt{15}}{6}$ into the second derivative. We can plug in $x=-1$. Then, we have

$$f''(-1) = 12(-1)^2 - 5 = 7 > 0,$$

so the function is concave up to the left of $x = -\frac{\sqrt{15}}{6}$.

To determine if the function is concave up or down on the interval to the right of $x = -\frac{\sqrt{15}}{6}$, or to determine if the second derivative of the function is positive or negative on the interval to the right of $x = -\frac{\sqrt{15}}{6}$, we need to plug any x value larger than $-\frac{\sqrt{15}}{6}$ and less than $\frac{\sqrt{15}}{6}$ into the second derivative because $x = \frac{\sqrt{15}}{6}$ is the next inflection point. We can plug in $x = 0$. Then, we have

$$f''(0) = 12(0)^2 - 5 = -5 < 0,$$

so the function is concave down to the right of $x = -\frac{\sqrt{15}}{6}$.

That also means that the function is concave down to the left of $x = \frac{\sqrt{15}}{6}$. To determine if the function is concave up or down on the interval to the right of $x = \frac{\sqrt{15}}{6}$, or to determine if the second derivative of the function is positive or negative on the interval to the right of $x = \frac{\sqrt{15}}{6}$, we need to plug any x value larger than $\frac{\sqrt{15}}{6}$ into the second derivative. We can plug in $x = 1$. Then, we have

$$f''(1) = 12(1)^2 - 5 = 7 > 0,$$

so the function is concave up to the right of $x = \frac{\sqrt{15}}{6}$.

Finally, the inflection points of the function are at $x = -\frac{\sqrt{15}}{6}$ (where the function changes from concave up to concave down) and $x = \frac{\sqrt{15}}{6}$ (where the function changes from concave down to concave up). The concave up intervals are $\left(-\infty, -\frac{\sqrt{15}}{6}\right)$ and $\left(\frac{\sqrt{15}}{6}, \infty\right)$, and the concave down interval is $\left(-\frac{\sqrt{15}}{6}, \frac{\sqrt{15}}{6}\right)$. If you want to visualize better, you can refer to the graph of $f(x) = x^4 - \frac{5}{2}x^2 + 1$ in figure 6.2.

Exercise Problems

1. Consider the function $f(x) = x^{\frac{1}{x}}$.
   a) Is it increasing or decreasing at $x = e^{-1}$?
   b) Is it increasing or decreasing at $x = 3$?
   c) Find its critical point(s). Determine if that/each of those critical point(s) is a maximum, a minimum, or neither.

2. Consider the function $f(x) = x^5 - x^3$.
   a) Find its inflection point(s).
   b) Find the intervals where the function is concave up or concave down.

Chapter 7

Indeterminate Forms of Limits and L'Hôpital's Rule

7.1 Indeterminate Forms and L'Hôpital's Rule

In this chapter, we will go back to limits, but we will not take a break from derivatives. Some limits have indeterminate forms, which are forms that do not give us a definite value (like 2, $\frac{1}{5}$, e, etc.), and we will have to do something with those forms to get a definite value for those limits. (Indeterminate means not able to determine.) Below is the list of indeterminate forms.

Indeterminate Forms

$$\frac{0}{0},\ \frac{\pm\infty}{\pm\infty},\ 0\times\pm\infty,\ (\pm\infty)^0,\ 0^0,\ 1^{\pm\infty},\ \infty-\infty$$

For the indeterminate forms $\frac{0}{0}$ and $\frac{\pm\infty}{\pm\infty}$, we can apply a rule called L'Hôpital's rule.

L'Hôpital's Rule
If

$$\lim_{x\to a}\frac{f(x)}{g(x)}=\frac{0}{0}\text{ or }\frac{\pm\infty}{\pm\infty}$$

and

$$\lim_{x\to a}\frac{f'(x)}{g'(x)}$$

exists, then

$$\lim_{x\to a}\frac{f(x)}{g(x)}=\lim_{x\to a}\frac{f'(x)}{g'(x)}.$$

Let us look at a few examples using L'Hôpital's rule.

Example 7.1-1: Find $\lim\limits_{x\to\infty}\dfrac{\ln x}{x}$.

Since

$$\lim_{x\to\infty}\ln x=\infty$$

and

$$\lim_{x\to\infty}x=\infty,$$

this is $\frac{\infty}{\infty}$ indeterminate form. So, we can apply L'Hôpital's rule. The derivative of the numerator is

$$\frac{d}{dx}(\ln x)=\frac{1}{x},$$

and the derivative of the denominator is

$$\frac{d}{dx}(x) = 1.$$

Applying the L'Hôpital's rule,

$$\lim_{x\to\infty} \frac{\ln x}{x} = \lim_{x\to\infty} \frac{\left(\frac{1}{x}\right)}{1} = \lim_{x\to\infty} \frac{1}{x} = 0.$$

Example 7.1-2: Find $\lim_{x\to 0} \dfrac{\arctan x}{x}$.
Since

$$\lim_{x\to 0} \arctan x = \arctan 0 = 0$$

and

$$\lim_{x\to 0} x = 0,$$

this is $\frac{0}{0}$ indeterminate form, so we can apply L'Hôpital's rule.
The derivative of the numerator is

$$\frac{d}{dx}(\arctan x) = \frac{1}{1+x^2},$$

and the derivative of the denominator is

$$\frac{d}{dx}(x) = 1.$$

Applying L'Hôpital's rule,

$$\lim_{x\to 0} \frac{\arctan x}{x} = \lim_{x\to 0} \frac{\left(\frac{1}{1+x^2}\right)}{1} = \lim_{x\to 0} \frac{1}{1+x^2} = 1.$$

For other indeterminate forms, we need to somehow re-arrange them to turn them into either $\frac{0}{0}$ or $\frac{\pm\infty}{\pm\infty}$ form so that we can apply L'Hôpital's rule. For $0 \times \pm\infty$, we usually would change it into either

$$\frac{\pm\infty}{\left(\frac{1}{0}\right)} = \frac{\pm\infty}{\pm\infty}$$

or

$$\frac{0}{\left(\frac{1}{\pm\infty}\right)} = \frac{0}{0}.$$

For the exponential indeterminate forms (i.e. $(\pm\infty)^0, 0^0$, and $1^{\pm\infty}$), we usually would take the natural logarithm (ln) of the expression to get $0 \times \pm\infty$ form and then arrange that $0 \times \pm\infty$ form to get either $\frac{0}{0}$ or $\frac{\pm\infty}{\pm\infty}$. For example, for $1^{\pm\infty}$, if we take natural logarithm, we would get

$$\ln\left(1^{\pm\infty}\right) = \pm\infty \times \ln(1) = \pm\infty \times 0.$$

For 0^0, if we take natural logarithm, we would get

$$\ln\left(0^0\right) = 0 \times \ln(0) = 0 \times -\infty.$$

For $(\pm\infty)^0$, if we take natural logarithm, we would get

$$\ln\left((\pm\infty)^0\right) = 0 \times \ln(\pm\infty) = 0 \times \pm\infty.$$

As we can see, there are nice general rules to approach each indeterminate form except for the $\infty - \infty$ form. For the $\infty - \infty$ form, we need to decide how to re-arrange the expression so that it changes into either $\frac{0}{0}$ or $\frac{\pm\infty}{\pm\infty}$ based on what the expression looks like. This is where algebraic skill is important in calculus.

Example 7.1-3: Find $\lim\limits_{x\to 0^+} x\ln x$.
Since

$$\lim_{x\to 0^+} x = 0$$

and

$$\lim_{x\to 0^+} \ln x = -\infty,$$

this is $0 \times (-\infty)$ indeterminate form. Re-arranging the expression, we get

$$\lim_{x\to 0^+} x\ln x = \lim_{x\to 0^+} \frac{\ln x}{\left(\frac{1}{x}\right)}.$$

Since

$$\lim_{x\to 0^+} \ln x = -\infty$$

and

$$\lim_{x\to 0^+} \frac{1}{x} = \infty,$$

this is $\frac{-\infty}{\infty}$ indeterminate form, so we can apply L'Hôpital's rule.
The derivative of the numerator is

$$\frac{d}{dx}(\ln x) = \frac{1}{x},$$

and the derivative of the denominator is

$$\frac{d}{dx}\left(\frac{1}{x}\right) = -\frac{1}{x^2}.$$

Applying L'Hôpital's rule,

$$\lim_{x\to 0^+} \frac{\ln x}{\left(\frac{1}{x}\right)} = \lim_{x\to 0^+} \frac{\left(\frac{1}{x}\right)}{\left(-\frac{1}{x^2}\right)} = \lim_{x\to 0^+} -x = 0.$$

Example 7.1-4: Find $\lim_{x \to 0^+} \ln x \ln(1-x)$.

Since

$$\lim_{x \to 0^+} \ln x = -\infty$$

and

$$\lim_{x \to 0^+} \ln(1-x) = \ln(1-0) = 0,$$

this is $0 \times (-\infty)$ indeterminate form. Re-arranging the expression, we get

$$\lim_{x \to 0^+} \ln x \ln(1-x) = \lim_{x \to 0^+} \frac{\ln(1-x)}{\left(\frac{1}{\ln x}\right)}.$$

Since

$$\lim_{x \to 0^+} \ln(1-x) = \ln(1-0) = 0$$

and

$$\lim_{x \to 0^+} \frac{1}{\ln x} = \frac{1}{\lim_{x \to 0^+} \ln x} = \frac{1}{-\infty} = 0,$$

this is $\frac{0}{0}$ indeterminate form, so we can apply L'Hôpital's rule. Applying the chain rule, the derivative of the numerator is

$$\frac{d}{dx}(\ln(1-x)) = -\frac{1}{1-x}.$$

Applying the quotient rule, the derivative of the denominator is

$$\frac{d}{dx}\left(\frac{1}{\ln x}\right) = -\frac{1}{x \ln^2 x}.$$

Applying L'Hôpital's rule,

$$\lim_{x \to 0^+} \frac{\ln(1-x)}{\left(\frac{1}{\ln x}\right)} = \lim_{x \to 0^+} \frac{\left(-\frac{1}{1-x}\right)}{\left(-\frac{1}{x \ln^2 x}\right)} = \lim_{x \to 0^+} \frac{x \ln^2 x}{1-x}. \tag{7.1}$$

Note that

$$\lim_{x \to 0^+} (1-x) = 1 - 0 = 1. \tag{7.2}$$

Now let us find

$$\lim_{x \to 0^+} x \ln^2 x.$$

Since

$$\lim_{x \to 0^+} x = 0$$

and

$$\lim_{x \to 0^+} \ln^2 x = (-\infty)^2 = \infty,$$

this is $0 \times \infty$ indeterminate form. Re-arranging the expression, we get

$$\lim_{x \to 0^+} x \ln^2 x = \lim_{x \to 0^+} \frac{\ln^2 x}{\left(\frac{1}{x}\right)}.$$

Since

$$\lim_{x \to 0^+} \ln^2 x = \infty$$

and

$$\lim_{x \to 0^+} \frac{1}{x} = \infty,$$

this is $\frac{\infty}{\infty}$ indeterminate form, so we can apply L'Hôpital's rule. Applying the chain rule, the derivative of the numerator is

$$\frac{d}{dx}\left(\ln^2 x\right) = \frac{2 \ln x}{x}.$$

The derivative of the denominator is

$$\frac{d}{dx}\left(\frac{1}{x}\right) = -\frac{1}{x^2}.$$

Applying L'Hôpital's rule,

$$\lim_{x \to 0^+} \frac{\ln^2 x}{\left(\frac{1}{x}\right)} = \lim_{x \to 0^+} \frac{\left(\frac{2 \ln x}{x}\right)}{\left(-\frac{1}{x^2}\right)} = \lim_{x \to 0^+} -2x \ln x = -2 \lim_{x \to 0^+} x \ln x.$$

In example 7.1-3, we found that

$$\lim_{x \to 0^+} x \ln x = 0,$$

so

$$\lim_{x \to 0^+} x \ln^2 x = \lim_{x \to 0^+} \frac{\ln^2 x}{\left(\frac{1}{x}\right)} = 0. \tag{7.3}$$

Plugging (7.2) and (7.3) into (7.1), our final answer is

$$\lim_{x \to 0^+} \ln x \ln (1 - x) = \lim_{x \to 0^+} \frac{\ln (1 - x)}{\left(\frac{1}{\ln x}\right)} = \lim_{x \to 0^+} \frac{x \ln^2 x}{1 - x} = \frac{0}{1} = 0.$$

Example 7.1-5: Find $\lim_{x \to 0^+} x^x$.
First, let

$$L = \lim_{x \to 0^+} x^x.$$

Since

$$\lim_{x \to 0^+} x = 0,$$

this is 0^0 indeterminate form. Taking natural logarithm of the expression, we get

$$\ln L = \lim_{x \to 0^+} \ln\left(x^x\right) = \lim_{x \to 0^+} x \ln x.$$

In example 7.1-3, we found that

$$\lim_{x \to 0^+} x \ln x = 0.$$

So,

$$\ln L = 0,$$

and

$$L = e^0 = 1.$$

Therefore, our final answer is

$$\lim_{x \to 0^+} x^x = 1.$$

Example 7.1-6: Find $\lim_{x \to \infty} x^{\frac{1}{x}}$.
First, let

$$L = \lim_{x \to \infty} x^{\frac{1}{x}}.$$

Since

$$\lim_{x \to \infty} x = \infty$$

and

$$\lim_{x \to \infty} \frac{1}{x} = 0,$$

this is ∞^0 indeterminate form. Taking natural logarithm of the expression, we get

$$\ln L = \lim_{x \to \infty} \ln\left(x^{\frac{1}{x}}\right) = \lim_{x \to \infty} \frac{\ln x}{x}.$$

In example 7.1-1, we found that

$$\lim_{x \to \infty} \frac{\ln x}{x} = 0.$$

So,

$$\ln L = 0,$$

and

$$L = e^0 = 1.$$

Therefore, our final answer is

$$\lim_{x \to \infty} x^{\frac{1}{x}} = 1.$$

Example 7.1-7: Find $\lim_{x\to 0} \sqrt[x]{1+\sin x}$.
Note that

$$\sqrt[x]{1+\sin x} = (1+\sin x)^{\frac{1}{x}}.$$

First, let

$$L = \lim_{x\to 0} (1+\sin x)^{\frac{1}{x}}.$$

Because

$$\lim_{x\to 0} \frac{1}{x}$$

does not exist, we need to split L into two limits:

$$L_1 = \lim_{x\to 0^+} (1+\sin x)^{\frac{1}{x}},$$

which is the limit as x approaches 0 from the right, and

$$L_2 = \lim_{x\to 0^-} (1+\sin x)^{\frac{1}{x}},$$

which is the limit as x approaches 0 from the left.
Now let us find L_1. Since

$$\lim_{x\to 0^+} \frac{1}{x} = \infty$$

and

$$\lim_{x\to 0^+} (1+\sin x) = 1+\sin 0 = 1,$$

this is 1^∞ indeterminate case. Taking natural logarithm of the expression, we get

$$\ln L_1 = \lim_{x\to 0^+} \ln\left((1+\sin x)^{\frac{1}{x}}\right) = \lim_{x\to 0^+} \frac{\ln(1+\sin x)}{x}.$$

Since

$$\lim_{x\to 0^+} \ln(1+\sin x) = \ln(1+\sin 0) = 0$$

and

$$\lim_{x\to 0^+} x = 0,$$

this is $\frac{0}{0}$ indeterminate case, so we can apply L'Hôpital's rule.
Applying the chain rule, the derivative of the numerator is

$$\frac{d}{dx}(\ln(1+\sin x)) = \frac{\cos x}{1+\sin x}.$$

The derivative of the denominator is

$$\frac{d}{dx}(x) = 1.$$

Applying L'Hôpital's rule,

$$\ln L_1 = \lim_{x\to 0^+} \frac{\left(\frac{\cos x}{1+\sin x}\right)}{1} = \lim_{x\to 0^+} \frac{\cos x}{1+\sin x} = \frac{\cos 0}{1+\sin 0} = 1.$$

So,

$$L_1 = e^1 = e. \tag{7.4}$$

Now let us find L_2. Since

$$\lim_{x\to 0^-} \frac{1}{x} = -\infty$$

and

$$\lim_{x\to 0^-} (1+\sin x) = 1+\sin 0 = 1,$$

this is $1^{-\infty}$ indeterminate case. Taking natural logarithm of the expression, we get

$$\ln L_2 = \lim_{x\to 0^-} \ln\left((1+\sin x)^{\frac{1}{x}}\right) = \lim_{x\to 0^-} \frac{\ln(1+\sin x)}{x}.$$

Since

$$\lim_{x\to 0^-} \ln(1+\sin x) = \ln(1+\sin 0) = 0$$

and

$$\lim_{x\to 0^-} x = 0,$$

this is $\frac{0}{0}$ indeterminate case, so we can apply L'Hôpital's rule. Applying L'Hôpital's rule,

$$\ln L_2 = \lim_{x\to 0^-} \frac{\left(\frac{\cos x}{1+\sin x}\right)}{1} = \lim_{x\to 0^-} \frac{\cos x}{1+\sin x} = \frac{\cos 0}{1+\sin 0} = 1.$$

So,

$$L_2 = e^1 = e. \tag{7.5}$$

From (7.4) and (7.5), we have

$$\lim_{x\to 0^+} (1+\sin x)^{\frac{1}{x}} = \lim_{x\to 0^-} (1+\sin x)^{\frac{1}{x}} = e.$$

Thus, our final answer is

$$\lim_{x\to 0} (1+\sin x)^{\frac{1}{x}} = \lim_{x\to 0^+} (1+\sin x)^{\frac{1}{x}} = \lim_{x\to 0^-} (1+\sin x)^{\frac{1}{x}} = e.$$

Example 7.1-8: Find $\lim\limits_{x\to 0^+} \left(\frac{1}{x} - \frac{1}{e^x - 1}\right)$.

Since

$$\lim_{x\to 0^+} \frac{1}{x} = \infty$$

and

$$\lim_{x\to 0^+} \frac{1}{e^x - 1} = \infty,$$

this is $\infty - \infty$ indeterminate form. We can re-arrange the expression by simply performing the subtraction of fractions:

$$\lim_{x\to 0^+} \left(\frac{1}{x} - \frac{1}{e^x - 1}\right) = \lim_{x\to 0^+} \frac{e^x - 1 - x}{xe^x - x}.$$

Since

$$\lim_{x\to 0^+} (e^x - 1 - x) = e^0 - 1 - 0 = 0$$

and

$$\lim_{x\to 0^+} (xe^x - x) = 0 \times e^0 - 0 = 0,$$

this is $\frac{0}{0}$ indeterminate form, so we can apply L'Hôpital's rule.
The derivative of the numerator is

$$\frac{d}{dx}(e^x - 1 - x) = e^x - 1.$$

Applying the product rule, the derivative of the denominator is

$$\frac{d}{dx}(xe^x - x) = xe^x + e^x - 1.$$

Applying L'Hôpital's rule,

$$\lim_{x\to 0^+} \frac{e^x - 1 - x}{xe^x - x} = \lim_{x\to 0^+} \frac{e^x - 1}{xe^x + e^x - 1}.$$

Since

$$\lim_{x\to 0^+} (e^x - 1) = e^0 - 1 = 0$$

and

$$\lim_{x\to 0^+} (xe^x + e^x - 1) = 0 \times e^0 + e^0 - 1 = 0,$$

this is $\frac{0}{0}$ indeterminate form, so we can apply L'Hôpital's rule again.
The derivative of the numerator is

$$\frac{d}{dx}(e^x - 1) = e^x.$$

Applying the product rule, the derivative of the denominator is

$$\frac{d}{dx}(xe^x + e^x - 1) = xe^x + 2e^x.$$

Applying L'Hôpital's rule,

$$\lim_{x\to 0^+} \frac{e^x - 1}{xe^x + e^x - 1} = \lim_{x\to 0^+} \frac{e^x}{xe^x + 2e^x} = \frac{e^0}{0 \times e^0 + 2e^0} = \frac{1}{2}.$$

Therefore, our final answer is

$$\lim_{x\to 0^+} \left(\frac{1}{x} - \frac{1}{e^x - 1}\right) = \frac{1}{2}.$$

Exercise Problems

1. Evaluate $\lim_{x\to 0^+} \ln x \ln(1+x)$.

2. Evaluate $\lim_{x\to 0^-} \left(\frac{1}{x} - \frac{1}{e^x - 1}\right)$.

3. Evaluate $\lim_{x\to 1} \left(\frac{1}{x-1} - \frac{1}{\ln x}\right)$.

4. Evaluate $\lim_{x\to\infty} x^2 e^{-x^2}$.

5. $\mathrm{erf}(x)$ is a function called error function. Given that $\lim_{x\to 0} \mathrm{erf}(x) = 0$, $\lim_{x\to\infty} \mathrm{erf}(x) = 1$, and $\frac{d}{dx}(\mathrm{erf}(x)) = \frac{2}{\sqrt{\pi}} e^{-x^2}$, evaluate:

   a) $\lim_{x\to 0} \frac{\mathrm{erf}(x)}{x}$,
   b) $\lim_{x\to\infty} (\mathrm{erf}(x))^x$.

Chapter 8

Stirling's Formula

8.1 Introduction to Factorial

Stirling's formula is a formula to approximate factorials which will be very useful in limits involving factorials. Before getting into Stirling's formula, we first need to know what a factorial is.

Factorial
For a positive integer n, n factorial, denoted by $n!$, is

$$n! = n(n-1)(n-2)\cdots 2\cdot 1.$$

For 0, $0! = 1$.

Basically, the factorial of a positive integer is the product of all positive integers less than or equal to that integer. For example, $3! = 3\times 2\times 1 = 6$, and $5! = 5\times 4\times 3\times 2\times 1 = 120$.

That is all you need to know about factorials for now. If you continue learning to more advanced calculus, you will learn that factorials can be extended to all complex numbers except negative integers through a function called gamma function. However, we will not go deep into gamma function in this book because it is beyond the scope of this book.

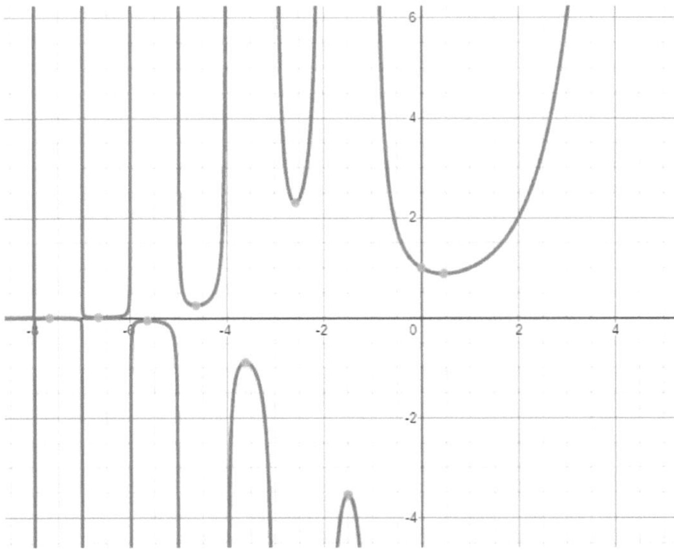

Figure 8.1: Graph of $f(x) = x!$ in the real domain

8.2 Stirling's Formula

Stirling's Formula
Stirling's formula (also known as Stirling's approximation) is given by

$$n! \sim \sqrt{2\pi n}\left(\frac{n}{e}\right)^n,$$

which means that the value of $\sqrt{2\pi n}\left(\frac{n}{e}\right)^n$ is approximately equal to the value of $n!$ for large n.

This formula is useful for limits involving $n!$ as n approaches infinity. Since the value of $\sqrt{2\pi n}\left(\frac{n}{e}\right)^n$ is approximately equal to the value of $n!$ for large n, it can be inferred that

$$\lim_{n\to\infty} n! = \lim_{n\to\infty} \sqrt{2\pi n}\left(\frac{n}{e}\right)^n, \tag{8.1}$$

i.e. the behavior of $n!$ as n approaches infinity is the same as the behavior of $\sqrt{2\pi n}\left(\frac{n}{e}\right)^n$ as n approaches infinity. So, we can substitute (8.1) into the limits involving $n!$ to make the limits easier to evaluate. Now let us take a look at some examples of limit problems involving Stirling's formula.

Example 8.2-1: Find $\lim\limits_{n\to\infty} \dfrac{n!}{n^n}$.
Substituting Stirling's formula,

$$\lim_{n\to\infty} \frac{n!}{n^n} = \lim_{n\to\infty} \frac{\sqrt{2\pi n}\left(\frac{n}{e}\right)^n}{n^n} = \lim_{n\to\infty} \frac{\sqrt{2\pi n}}{e^n}.$$

Since

$$\lim_{n\to\infty} \sqrt{2\pi n} = \infty$$

and

$$\lim_{n\to\infty} e^n = \infty,$$

this is $\frac{\infty}{\infty}$ indeterminate form, so we can apply L'Hôpital's rule.
The derivative of the numerator is

$$\frac{d}{dn}\left(\sqrt{2\pi n}\right) = \sqrt{\frac{\pi}{2n}},$$

and the derivative of the denominator is

$$\frac{d}{dn}\left(e^n\right) = e^n.$$

Applying L'Hôpital's rule,

$$\lim_{n\to\infty} \frac{\sqrt{2\pi n}}{e^n} = \lim_{n\to\infty} \frac{\sqrt{\frac{\pi}{2n}}}{e^n} = \sqrt{\frac{\pi}{2}} \lim_{n\to\infty} \frac{1}{\sqrt{n}e^n} = \sqrt{\frac{\pi}{2}} \times \frac{1}{\infty} = 0.$$

Therefore, our final answer is

$$\lim_{n\to\infty} \frac{n!}{n^n} = 0.$$

Example 8.2-2: Find $\lim_{n\to\infty} \frac{n}{\sqrt[n]{n!}}$.
Substituting Stirling's formula,

$$\lim_{n\to\infty} \frac{n}{\sqrt[n]{n!}} = \lim_{n\to\infty} \frac{n}{\sqrt[n]{\sqrt{2\pi n}\left(\frac{n}{e}\right)^n}} = \lim_{n\to\infty} \frac{e}{\sqrt[2n]{2\pi n}}. \tag{8.2}$$

Now let

$$L = \lim_{n\to\infty} \sqrt[2n]{2\pi n} = \lim_{n\to\infty} (2\pi n)^{\frac{1}{2n}},$$

and let us evaluate L. Since

$$\lim_{n\to\infty} 2\pi n = \infty$$

and

$$\lim_{n\to\infty} \frac{1}{2n} = \frac{1}{\infty} = 0,$$

this is ∞^0 indeterminate form. Taking natural logarithm of the expression, we get

$$\ln L = \lim_{n\to\infty} \ln\left((2\pi n)^{\frac{1}{2n}}\right) = \lim_{n\to\infty} \frac{\ln(2\pi n)}{2n}.$$

Since

$$\lim_{n\to\infty} \ln(2\pi n) = \ln(\infty) = \infty$$

and

$$\lim_{n\to\infty} 2n = \infty,$$

this is $\frac{\infty}{\infty}$ indeterminate form, so we can apply L'Hôpital's rule.
The derivative of the numerator is

$$\frac{d}{dn}(\ln(2\pi n)) = \frac{d}{dn}(\ln(2\pi) + \ln(n)) = \frac{1}{n},$$

and the derivative of the denominator is

$$\frac{d}{dn}(2n) = 2.$$

Applying L'Hôpital's rule,

$$\lim_{n\to\infty} \frac{\ln(2\pi n)}{2n} = \lim_{n\to\infty} \frac{\left(\frac{1}{n}\right)}{2} = \lim_{n\to\infty} \frac{1}{2n} = \frac{1}{\infty} = 0.$$

Thus, we have $\ln L = 0$. So,

$$L = \lim_{n\to\infty} \sqrt[2n]{2\pi n} = e^0 = 1.$$

Plugging this into (8.2), our final answer is

$$\lim_{n\to\infty} \frac{n}{\sqrt[n]{n!}} = \lim_{n\to\infty} \frac{e}{\sqrt[2n]{2\pi n}} = \frac{e}{1} = e.$$

Exercise Problems

1. 4!=?

2. Evaluate $\lim_{n\to\infty} \frac{\ln(n!)}{n^2}$.

3. Evaluate $\lim_{n\to\infty} \left(n - \left(n + \frac{1}{2}\right)\ln(n) + \ln(n!)\right)$.

Chapter 9

Floor Function, Ceiling Function, and Fractional Part Function

9.1 Floor Function

Let us start with the definition of the floor function - also known as the greatest integer function.

Floor Function (Greatest Integer Function)
By definition, the floor function of a number x, denoted by $\lfloor x \rfloor$, is the largest integer smaller than or equal to x. For example, $\lfloor 2 \rfloor = 2$, and $\lfloor 3.4 \rfloor = 3$.

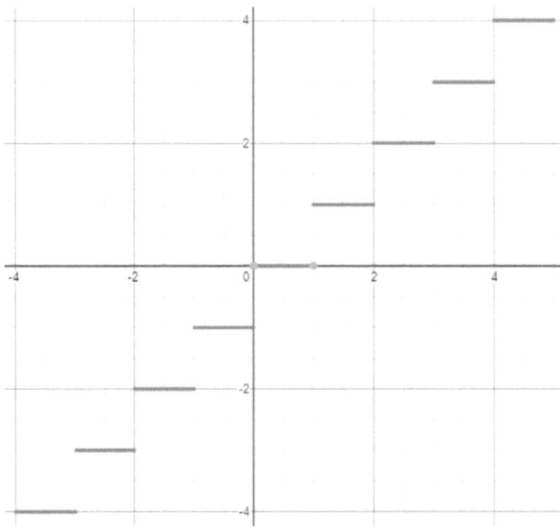

Figure 9.1: Graph of $f(x) = \lfloor x \rfloor$

Above is the graph of $f(x) = \lfloor x \rfloor$. There should be open circles at the points $(n+1, n)$ where n is an integer, i.e. the points (1,0), (2,1), (3,2), etc., but Desmos does not show the open circles unless you hold the mouse over those points. If we look at the graph, we can see that $\lfloor x \rfloor = n$ for $x \in [n, n+1)$ where n is an integer. For example for $x \in [1, 2)$, $\lfloor x \rfloor = 1$. Also, note that the domain of the floor function is all real numbers, and the range is the set of integers.

Recall that the derivative of a function is the slope of the tangent line to the curve of that function. From observing the graph of $f(x) = \lfloor x \rfloor$, we can conclude that the derivative of $\lfloor x \rfloor$ is undefined at integer x values because the graph is discontinuous at those points and is 0 at other x values.

Derivative of $\lfloor \mathbf{x} \rfloor$

$$\frac{d}{dx}(\lfloor x \rfloor) = \begin{cases} \text{undefined} & \text{if } x \in \mathbb{Z} \\ 0 & \text{otherwise} \end{cases}$$

The notation "$\mathbb{Z}$" means the set of integers, and the notation "$\in$" means being in some set. So, "$x \in \mathbb{Z}$" means x is in the set of integers, or x is an integer.

Our observation of the derivative of $\lfloor x \rfloor$ can be extended to $\lfloor f(x) \rfloor$ for any $f(x)$ a function of x. In general, we have the following rule for the derivative of $\lfloor f(x) \rfloor$.

Derivative of $\lfloor \mathbf{f(x)} \rfloor$

$$\frac{d}{dx}(\lfloor f(x) \rfloor) = \begin{cases} \text{undefined} & \text{if } x \in \{x \mid \lfloor f(x) \rfloor \text{ is discontinuous}\} \\ 0 & \text{otherwise} \end{cases}$$

The notation "$\mid$" means "such that". So, "$\{x \mid \lfloor f(x) \rfloor$ is discontinuous$\}$" means the set of x such that $\lfloor f(x) \rfloor$ is discontinuous. Just like how the derivative of $\lfloor x \rfloor$ is undefined at integer x values because $\lfloor x \rfloor$ is discontinuous at those points, the derivative of $\lfloor f(x) \rfloor$ is undefined at the points where $\lfloor f(x) \rfloor$ is discontinuous.

So, how do we know where $\lfloor f(x) \rfloor$ is continuous or discontinuous? Well, just like how $\lfloor x \rfloor$ is discontinuous at integer x values, in general, $\lfloor f(x) \rfloor$ is discontinuous at x values where $f(x)$ is an integer. However, there are a few exceptions to that general rule.

Continuity of $\lfloor \mathbf{f(x)} \rfloor$
Usually, $\lfloor f(x) \rfloor$ is discontinuous at $x = x_0$ if $f(x_0)$ is an integer or $f(x)$ is discontinuous at $x = x_0$, and $\lfloor f(x) \rfloor$ is continuous at $x = x_0$ if $f(x_0)$ is not an integer and $f(x)$ is continuous at $x = x_0$. However, there are a few exceptions:

- If $f(x)$ has a minimum at $x = x_0$, then $\lfloor f(x) \rfloor$ is continuous at $x = x_0$, even if $f(x_0)$ is an integer.

- If $f(x)$ is a constant function, then $\lfloor f(x) \rfloor$ is also a constant function, and $\lfloor f(x) \rfloor$ is continuous for all real x.

• If $f(x)$ is a non-constant function whose outputs (or range) are integers only, like the floor function, then $\lfloor f(x) \rfloor = f(x)$, and $\lfloor f(x) \rfloor$ is discontinuous wherever $f(x)$ is discontinuous.

• If $f(x)$ is a function whose range is $[n, n+1)$ or shorter, where n is an integer, then $\lfloor f(x) \rfloor = n$, which is constant, and $\lfloor f(x) \rfloor$ is continuous for all real x.

For the first exception, we can take a look at certain points of some graphs as examples: the point at $x = 0$ of $\lfloor x^2 \rfloor$, the point at $x = 1$ of $\lfloor |x-1| \rfloor$, and the point at $x = -\frac{\pi}{2}$ of $\lfloor \sin x \rfloor$. However, I will not show them to you here, and I will leave that as an exercise for you to observe the graphs on your own. You can use Desmos or whatever graphing utility you have to graph and observe the graphs I listed above. Also, as a reminder, we learned how to find minima of a function in chapter 6.

The second exception is trivial. If $f(x)$ is always the same number, then $\lfloor f(x) \rfloor$ must always be the same number as well.

As for the third exception, we need to pay attention to the definition of the floor function. From the definition, we can see that $\lfloor x \rfloor = x$ if x is an integer. So, if $f(x)$ is always an integer, then $\lfloor f(x) \rfloor = f(x)$.

For the fourth exception, we need to pay attention to the graph of the floor function. From the graph, we observed that $\lfloor x \rfloor = n$ for $x \in [n, n+1)$ where n is an integer. So, if the range of $f(x)$ is $[n, n+1)$ or shorter, i.e. $f(x) \in [n, n+1)$, then $\lfloor f(x) \rfloor = n$.

There are still some final notes that need to be made about the continuity of $\lfloor f(x) \rfloor$. If $f(x)$ is a piece-wise function, meaning different parts of $f(x)$ are defined differently, then we apply the rules for the continuity of $\lfloor f(x) \rfloor$ to each different part that are defined differently. Then, we take the floor function of the value at the point where the definition of $f(x)$ changes to see if $\lfloor f(x) \rfloor$ is continuous at that point. Also, what if $f(x)$ is a function whose range is $[n, n+2)$ where n is an integer? Then, we can split $f(x)$ into two parts where $f(x) \in [n, n+1)$ and $f(x) \in [n+1, n+2)$, and we can apply the fourth exception rule to each part separately.

Overall, the most important thing in determining whether $\lfloor f(x) \rfloor$ is continuous at some point is observing the behavior of $f(x)$ and using what we know about the floor function. The general rule and exceptions I listed only serve as general guidelines for most functions - not all. If you see some "weird" function, like the one whose range is $[n, n+2)$, then you will have to observe the function and somehow manipulate it to be able to apply the rules I listed.

9.2 Ceiling Function

Now let us move on to the discussion of the ceiling function, which is very similar to the floor function. So, in this section, there will be less explanations in words because everything is similar to the last section about the floor function.

Ceiling Function
By definition, the ceiling function of a number x, denoted by $\lceil x \rceil$, is the smallest integer larger than or equal to x. For example, $\lceil 3 \rceil = 3$, and $\lceil 5.3 \rceil = 6$.

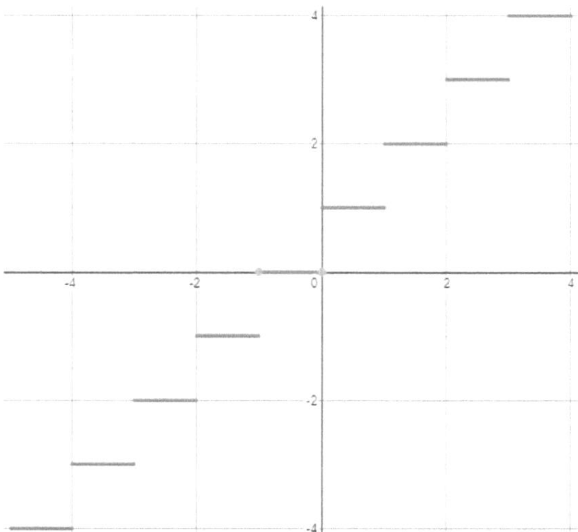

Figure 9.2: Graph of $f(x) = \lceil x \rceil$

Above is the graph of $f(x) = \lceil x \rceil$. There should be open circles at the points $(n, n+1)$ where n is an integer, i.e. the points (0,1), (1,2), (2,3), etc. If we look at the graph, we can see that $\lceil x \rceil = n+1$ for $x \in (n, n+1]$ where n is an integer. For example for $x \in (2,3]$, $\lceil x \rceil = 3$. Also, note that the domain of the ceiling function is all real numbers, and the range is the set of integers.

Similarly to the derivative of the floor function, we can derive the following piece-wise definition for the derivative of the ceiling function.

Derivative of $\lceil \mathbf{x} \rceil$

$$\frac{d}{dx}\left(\lceil x \rceil\right) = \begin{cases} \text{undefined} & \text{if } x \in \mathbb{Z} \\ 0 & \text{otherwise} \end{cases}$$

We can also extend this to the derivative of $\lceil f(x) \rceil$ for any $f(x)$ a function of x.

Derivative of $\lceil \mathbf{f(x)} \rceil$

$$\frac{d}{dx}\left(\lceil f(x) \rceil\right) = \begin{cases} \text{undefined} & \text{if } x \in \{x \mid \lceil f(x) \rceil \text{ is discontinuous}\} \\ 0 & \text{otherwise} \end{cases}$$

The following are the general "guidelines" to determine where $\lceil f(x) \rceil$ is discontinuous or continuous. You should be a little careful here because the rules for the continuity of $\lceil f(x) \rceil$ are not exactly the same as the rules for the continuity of $\lfloor f(x) \rfloor$.

Continuity of $\lceil \mathbf{f(x)} \rceil$
Usually, $\lceil f(x) \rceil$ is discontinuous at $x = x_0$ if $f(x_0)$ is an integer or $f(x)$ is discontinuous at $x = x_0$, and $\lceil f(x) \rceil$ is continuous at $x = x_0$ if $f(x_0)$ is not an integer and $f(x)$ is continuous at $x = x_0$. However, there are a few exceptions:

- If $f(x)$ has a maximum at $x = x_0$, then $\lceil f(x) \rceil$ is continuous at $x = x_0$, even if $f(x_0)$ is an integer.

- If $f(x)$ is a constant function, then $\lceil f(x) \rceil$ is also a constant function, and $\lceil f(x) \rceil$ is continuous for all real x.

- If $f(x)$ is a non-constant function whose outputs (or range) are integers only, like the floor function and the ceiling function, then $\lceil f(x) \rceil = f(x)$, and $\lceil f(x) \rceil$ is discontinuous wherever $f(x)$ is discontinuous.

- If $f(x)$ is a function whose range is $(n, n+1]$ or shorter, where n is an integer, then $\lceil f(x) \rceil = n+1$, which is constant, and $\lceil f(x) \rceil$ is continuous for all real x.

For the first exception, we can look at certain points of some graphs as examples: the point at $x = 0$ of $\lceil \cos x \rceil$, the point at $x = -1$ of $\left\lceil -(x+1)^2 \right\rceil$, and the point at $x = 0$ of $\lceil x^3 - x^2 \rceil$. Also, as a reminder, we learned how to find maxima of a function in chapter 6.

The second exception is trivial. The ceiling function of a number is obviously always the same number.

For the third exception, we need to pay attention to the definition of the ceiling function. From the definition of the ceiling function, we can see that $\lceil x \rceil = x$ if x is an integer. So, if $f(x)$ is always an integer, then $\lceil f(x) \rceil = f(x)$.

For the fourth exception, we need to pay attention to the graph of the ceiling function. From the graph, we observed that $\lceil x \rceil = n+1$ for $x \in (n, n+1]$ where n is an integer. So, if the range of $f(x)$ is $(n, n+1]$ or shorter, i.e. $f(x) \in (n, n+1]$, then $\lceil f(x) \rceil = n+1$.

Again, these rules only serve as general guidelines for most functions, not all. The most important thing in determining where the graph of $\lceil f(x) \rceil$ is continuous or discontinuous is observing the graph of $f(x)$ and somehow manipulating it if necessary.

9.3 Fractional Part Function

Now let us move on to the fractional part function, and let us start with the definition (or definitions, actually) of the fractional part function, denoted by $\{x\}$.

Fractional Part Function
Definition 1:

$$\{x\} = x - \lfloor x \rfloor, \forall x \in \mathbb{R}.$$

Definition 2:

$$\{x\} = \begin{cases} x - \lfloor x \rfloor & \text{if } x \geq 0 \\ x - \lceil x \rceil & \text{if } x < 0. \end{cases}$$

The notation "$\forall$" means "for all", and "$\mathbb{R}$" is the set of real numbers. So, "$\forall x \in \mathbb{R}$" means "for all real x". Unfortunately, as you can see, there is not a unified definition of the fractional part function for negative real numbers. However, that does not really matter because we usually would deal with the fractional part function of positive numbers more often than the fractional part function of negative numbers.

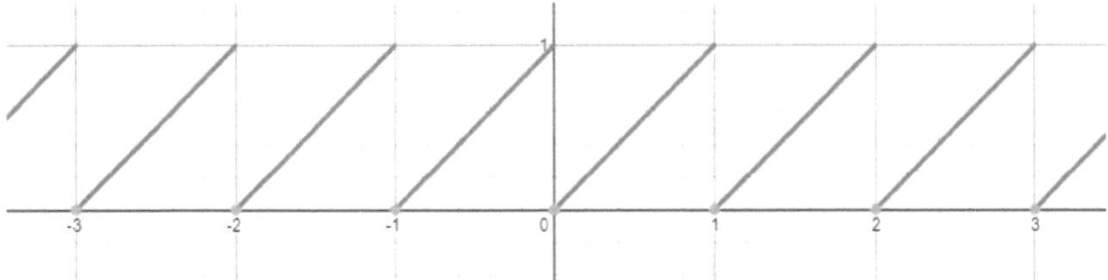

Figure 9.3: Graph of $f(x) = \{x\}$ using definition 1

Using the definition 1, we obtain the graph of the fractional part function as in figure 9.3. There should be open circles at the points $(n, 1)$ where n is an integer, i.e. the points (0,1), (1,1), (2,1), etc. The domain is all real numbers, and the range is [0,1).

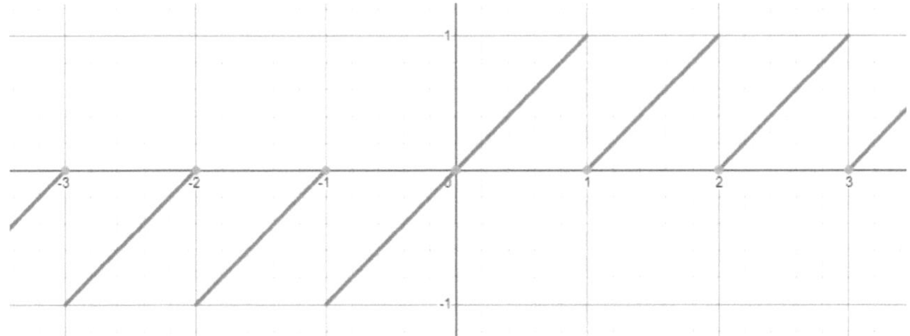

Figure 9.4: Graph of $f(x) = \{x\}$ using definition 2

Using the definition 2, we obtain the graph of the fractional part function as in figure 9.4. There should be open circles at the points $(n, 1)$ and $(-n, -1)$ where n is a positive integer, i.e. the points (-2,-1), (-1,-1), (1,1), (2,1), etc. The domain is all real numbers, and the range is (-1,1).

Since the fractional part function is defined using the floor function and the ceiling function, there is no need to discuss the derivative of the fractional part function.

Exercise Problems

1. $\lfloor \pi \rfloor - \lceil e \rceil = ?$

2. $\lceil \pi \rceil - \lfloor e \rfloor = ?$

3. $\{4.567\} = ?$

4. Using the definition 1 of the fractional part function, simplify the following expression: $\{\{x\}\}$.

5.

> **Iterated Function**
> By definition, n-th iterate of a function $f(x)$ is
> $$f^n(x) = \underbrace{(f \circ f \circ f \circ \cdots \circ f)}_{n \text{ times}}(x),$$
> and
> $$f^n(x) = \left(f \circ f^{n-1}\right)(x).$$

a) Let $f(x) = \lfloor x \rfloor$. Prove that $f^n(x) = f(x)$ for any positive integer n.
b) Let $f(x) = \lceil x \rceil$. Prove that $f^n(x) = f(x)$ for any positive integer n.

6.

> **Swinging Factorial**
> By definition, for any positive integer n, the swinging factorial of n, denoted by $n\wr$, is
> $$n\wr = \frac{n!}{(\lfloor n/2 \rfloor !)^2}.$$

a) $5\wr =?$
b) $6\wr =?$

Chapter 10

Some Basic Higher Derivatives

10.1 Higher Than Second Derivative

So far, we have learned first derivative and second derivative. There are also higher derivatives such as third derivative, fourth derivative, etc. Each higher derivative is the derivative of the lower derivative by one order. For example, third derivative is the derivative of second derivative, and fourth derivative is the derivative of third derivative. Below is the notation for n-th derivative where n is a positive integer.

> **Notation of Higher Derivatives**
> For any n a positive integer, n-th derivative of a function $f(x)$ is denoted by
> $$\frac{d^n}{dx^n}(f(x)) = f^{(n)}(x).$$

Example 10.1-1: Find the third derivative of $x \ln x$.
Applying the product rule, the first derivative of $x \ln x$ is

$$\frac{d}{dx}(x \ln x) = \ln(x) + 1. \tag{10.1}$$

Differentiating (10.1) with respect to x, the second derivative is

$$\frac{d^2}{dx^2}(x \ln x) = \frac{1}{x}. \tag{10.2}$$

Differentiating (10.2) with respect to x, the third derivative is

$$\frac{d^3}{dx^3}(x \ln x) = -\frac{1}{x^2}.$$

Example 10.1-2: For $f(x) = \ln(\cos x)$, find $f^{(5)}\left(\frac{\pi}{4}\right)$.
Applying the chain rule, the first derivative of $f(x)$ is

$$f^{(1)}(x) = \frac{d}{dx}(\ln(\cos x)) = -\tan x. \tag{10.3}$$

Differentiating (10.3) with respect to x, the second derivative of $f(x)$ is

$$f^{(2)}(x) = -\sec^2 x. \tag{10.4}$$

Differentiating (10.4) with respect to x by applying the chain rule, the third derivative of $f(x)$ is

$$f^{(3)}(x) = -2\sec^2 x \tan x. \tag{10.5}$$

Differentiating (10.5) with respect to x by applying the product rule and the chain rule, the fourth derivative of $f(x)$ is

$$f^{(4)}(x) = -4\sec^2 x \tan^2 x - 2\sec^4 x. \tag{10.6}$$

Differentiating (10.6) with respect to x by applying the product rule and the chain rule, the fifth derivative of $f(x)$ is

$$f^{(5)}(x) = -8\sec^2 x \tan^3 x - 16\sec^4 x \tan x.$$

Plugging in $x = \frac{\pi}{4}$, our final answer is

$$f^{(5)}\left(\frac{\pi}{4}\right) = -8\sec^2 \frac{\pi}{4} \tan^3 \frac{\pi}{4} - 16\sec^4 \frac{\pi}{4} \tan \frac{\pi}{4} = -80.$$

10.2 Some General Patterns for Higher Derivatives

Here in this section, we will discuss the general patterns for higher derivatives of some types of function. First, let us take a look at the positive powers of x. For any positive real a, if we keep differentiating x^a, we would get

$$\begin{aligned}
\frac{d}{dx}\left(x^a\right) &= ax^{a-1}, \\
\frac{d^2}{dx^2}\left(x^a\right) &= a(a-1)x^{a-2}, \\
\frac{d^3}{dx^3}\left(x^a\right) &= a(a-1)(a-2)x^{a-3}, \\
\frac{d^4}{dx^4}\left(x^a\right) &= a(a-1)(a-2)(a-3)x^{a-4},
\end{aligned}$$

and so on. So, we can see the general pattern

$$\frac{d^n}{dx^n}\left(x^a\right) = \frac{a!\, x^{a-n}}{(a-n)!}$$

for any positive real a. Now let us take a look at the negative powers of x. For any positive real a, if we keep differentiating x^{-a}, we would get

$$\begin{aligned}
\frac{d}{dx}\left(x^{-a}\right) &= (-a)x^{-a-1}, \\
\frac{d^2}{dx^2}\left(x^{-a}\right) &= (-a)(-a-1)x^{-a-2}, \\
\frac{d^3}{dx^3}\left(x^{-a}\right) &= (-a)(-a-1)(-a-2)x^{-a-3}, \\
\frac{d^4}{dx^4}\left(x^{-a}\right) &= (-a)(-a-1)(-a-2)(-a-3)x^{-a-4},
\end{aligned}$$

and so on. So, we can see the general pattern

$$\frac{d^n}{dx^n}\left(x^{-a}\right) = \frac{(-1)^n (a+n-1)!\, x^{-a-n}}{(a-1)!} = \frac{(-1)^n (a+n-1)!}{(a-1)!\, x^{a+n}}$$

for any positive real a.

n-th Derivatives of Powers of x

For any positive real a and positive integer n,

$$\frac{d^n}{dx^n}\left(x^{a}\right) = \frac{a!\, x^{a-n}}{(a-n)!},$$

and

$$\frac{d^n}{dx^n}\left(x^{-a}\right) = \frac{(-1)^n (a+n-1)!}{(a-1)!\, x^{a+n}}.$$

Next let us look at the exponents of x. For any non-zero complex number a, if we keep differentiating a^x, we would get

$$\frac{d}{dx}\left(a^x\right) = a^x \ln a,$$
$$\frac{d^2}{dx^2}\left(a^x\right) = a^x \ln^2 a,$$
$$\frac{d^3}{dx^3}\left(a^x\right) = a^x \ln^3 a,$$
$$\frac{d^4}{dx^4}\left(a^x\right) = a^x \ln^4 a,$$

and so on. So, we can see the general pattern

$$\frac{d^n}{dx^n}\left(a^x\right) = a^x \ln^n a$$

for any non-zero complex number a.

n-th Derivatives of Exponents of x

For any non-zero complex number a and positive integer n,

$$\frac{d^n}{dx^n}\left(a^x\right) = a^x \ln^n a.$$

Now let us look at the logarithms of x. For any non-zero complex number a, if we keep differentiating $\log_a x$, we would get

$$\begin{aligned}\frac{d}{dx}(\log_a x) &= \frac{1}{x \ln a},\\ \frac{d^2}{dx^2}(\log_a x) &= -\frac{1}{x^2 \ln a},\\ \frac{d^3}{dx^3}(\log_a x) &= \frac{2}{x^3 \ln a},\\ \frac{d^4}{dx^4}(\log_a x) &= -\frac{2 \times 3}{x^4 \ln a},\end{aligned}$$

and so on. So, we can see the general pattern

$$\frac{d^n}{dx^n}(\log_a x) = \frac{(-1)^{n+1}(n-1)!}{x^n \ln a}$$

for any non-zero complex number a.

n-th Derivatives of Logarithms of x

For any non-zero complex number a and positive integer n,

$$\frac{d^n}{dx^n}(\log_a x) = \frac{(-1)^{n+1}(n-1)!}{x^n \ln a}.$$

Finally, let us look at the general pattern for higher derivatives of sine and cosine. If we keep differentiating $\sin x$, we would get

$$\begin{aligned}\frac{d}{dx}(\sin x) &= \cos x,\\ \frac{d^2}{dx^2}(\sin x) &= -\sin x,\\ \frac{d^3}{dx^3}(\sin x) &= -\cos x,\\ \frac{d^4}{dx^4}(\sin x) &= \sin x,\end{aligned}$$

and the cycle repeats for every four times of differentiation. Also, note that

$$\begin{aligned}
\sin\left(x+\frac{1\pi}{2}\right) &= \sin x\cos\frac{1\pi}{2}+\cos x\sin\frac{1\pi}{2} = \cos x,\\
\sin\left(x+\frac{2\pi}{2}\right) &= \sin x\cos\frac{2\pi}{2}+\cos x\sin\frac{2\pi}{2} = -\sin x,\\
\sin\left(x+\frac{3\pi}{2}\right) &= \sin x\cos\frac{3\pi}{2}+\cos x\sin\frac{3\pi}{2} = -\cos x,\\
\sin\left(x+\frac{4\pi}{2}\right) &= \sin x\cos\frac{4\pi}{2}+\cos x\sin\frac{4\pi}{2} = \sin x,
\end{aligned}$$

and the cycle repeats because $\sin x$ is a periodic function with a period of 2π. So, we have the general pattern

$$\frac{d^n}{dx^n}(\sin x) = \sin\left(x+\frac{n\pi}{2}\right).$$

Also, by applying the chain rule, we can see that

$$\frac{d^n}{dx^n}(\sin(ax)) = a^n\sin\left(ax+\frac{n\pi}{2}\right)$$

for any complex number a. Now let us take a look at cosine. If we keep differentiating $\cos x$, we would get

$$\begin{aligned}
\frac{d}{dx}(\cos x) &= -\sin x,\\
\frac{d^2}{dx^2}(\cos x) &= -\cos x,\\
\frac{d^3}{dx^3}(\cos x) &= \sin x,\\
\frac{d^4}{dx^4}(\cos x) &= \cos x,
\end{aligned}$$

and the cycle repeats for every four times of differentiation. Also, note that

$$\begin{aligned}
\cos\left(x+\frac{1\pi}{2}\right) &= \cos x\cos\frac{1\pi}{2}-\sin x\sin\frac{1\pi}{2} = -\sin x,\\
\cos\left(x+\frac{2\pi}{2}\right) &= \cos x\cos\frac{2\pi}{2}-\sin x\sin\frac{2\pi}{2} = -\cos x,\\
\cos\left(x+\frac{3\pi}{2}\right) &= \cos x\cos\frac{3\pi}{2}-\sin x\sin\frac{3\pi}{2} = \sin x,\\
\cos\left(x+\frac{4\pi}{2}\right) &= \cos x\cos\frac{4\pi}{2}-\sin x\sin\frac{4\pi}{2} = \cos x,
\end{aligned}$$

and the cycle repeats because $\cos x$ is a periodic function with a period of 2π. So, we have the general pattern

$$\frac{d^n}{dx^n}(\cos x) = \cos\left(x + \frac{n\pi}{2}\right).$$

Also, by applying the chain rule, we can see that

$$\frac{d^n}{dx^n}(\cos(ax)) = a^n \cos\left(ax + \frac{n\pi}{2}\right)$$

for any complex number a.

n-th Derivatives of Sine and Cosine
For any complex number a and positive integer n,

$$\frac{d^n}{dx^n}(\sin(ax)) = a^n \sin\left(ax + \frac{n\pi}{2}\right),$$

and

$$\frac{d^n}{dx^n}(\cos(ax)) = a^n \cos\left(ax + \frac{n\pi}{2}\right).$$

To summarize, in this chapter we have learned about n-th derivatives where n is a positive integer. However, n does not have to be a positive integer, and it can be any positive real number or complex number. There can be 1/2-th derivative or even i-th derivative (where $i = \sqrt{-1}$). Of course, we are not going to discuss those types of derivatives because they are beyond the scope of this book. This field of study is called fractional derivatives, and you can study fractional derivatives once you have a good understanding of how factorials, logarithms, and exponents can be extended to complex numbers.

Exercise Problems

1. Let $f(x) = x^2 e^x$. Find $f^{(4)}(2)$.

2. Let $f(x) = x^x$. Find $f^{(3)}(1)$.

3. Find $\dfrac{d^n}{dx^n}(\sin(5x)\cos(3x))$ where n is a positive integer. (Hint: use a trigonometric identity)

4. Find the general pattern for $\dfrac{d^n}{dx^n}(xe^x)$ where n is a positive integer.

5. If $f(x) = e^x + \cos(x)$, find the value of $f^{(1)}(0) + f^{(2)}(0) + f^{(3)}(0) + f^{(4)}(0) + \cdots + f^{(124)}(0)$.

Chapter 11

Introduction to Summation, Product, Arithmetic Sequence, and Geometric Sequence

11.1 Introduction to Summation

In this section, we will learn about the summation, which is a sum of an expression from some value to some other value.

Summation
Let $f(n)$ be an expression in terms of n. Then, the summation of $f(n)$ from $n = a$ to $n = b$, where a and b are integers and $a \leq b$, is

$$\sum_{n=a}^{b} f(n) = f(a) + f(a+1) + f(a+2) + \cdots + f(b).$$

n is called the index of the summation. $n = a$ is the lower limit, and $n = b$ is the upper limit of the summation.

Now let us look at some examples.

Example 11.1-1: Find $\sum_{n=1}^{4} n^2$.

$$\sum_{n=1}^{4} n^2 = 1^2 + 2^2 + 3^2 + 4^2 = 30.$$

Example 11.1-2: Find $\sum_{n=3}^{7} \left(n + \frac{1}{n}\right)$.

$$\begin{aligned}\sum_{n=3}^{7} \left(n + \frac{1}{n}\right) &= \left(3 + \frac{1}{3}\right) + \left(4 + \frac{1}{4}\right) + \left(5 + \frac{1}{5}\right) + \left(6 + \frac{1}{6}\right) + \left(7 + \frac{1}{7}\right) \\ &= 26\frac{13}{140}.\end{aligned}$$

Example 11.1-3: Find $\sum_{n=1}^{8} n^3$.

$$\sum_{n=1}^{8} n^3 = 1^3 + 2^3 + 3^3 + 4^3 + 5^3 + 6^3 + 7^3 + 8^3 = 1296.$$

Example 11.1-4: Find $\sum_{n=1}^{10} n$.

$$\sum_{n=1}^{10} n = 1 + 2 + 3 + 4 + 5 + 6 + 7 + 8 + 9 + 10 = 55.$$

Note that the summation has the following obvious but important properties.

Properties of Summation
Let $f(n)$ and $g(n)$ be expressions in terms of n. Then,

$$\sum_{n=a}^{b}(f(n)+g(n))=\sum_{n=a}^{b}f(n)+\sum_{n=a}^{b}g(n).$$

If $b \geq a \geq c$ for some integers a, b, and c, and $f(n)$ is an expression in terms of n, then

$$\sum_{n=a}^{b}f(n)=\sum_{n=c}^{b}f(n)-\sum_{n=c}^{a}f(n).$$

If $b \geq a$ for some integers a and b, c is an integer, and $f(n)$ is an expression in terms of n, then we can re-index a summation, or change the limits of the index of a summation, as follows:

$$\sum_{n=a}^{b}f(n)=\sum_{n=a+c}^{b+c}f(n-c).$$

For example, the summation in example 11.1-2 can be rewritten as

$$\sum_{n=3}^{7}\left(n+\frac{1}{n}\right)=\sum_{n=3}^{7}n+\sum_{n=3}^{7}\frac{1}{n},$$

$$\sum_{n=3}^{7}\left(n+\frac{1}{n}\right)=\sum_{n=1}^{7}\left(n+\frac{1}{n}\right)-\sum_{n=1}^{3}\left(n+\frac{1}{n}\right),$$

or

$$\sum_{n=3}^{7}\left(n+\frac{1}{n}\right)=\sum_{n=4}^{8}\left(n-1+\frac{1}{n-1}\right)=\sum_{n=2}^{6}\left(n+1+\frac{1}{n+1}\right).$$

Also, there are some useful formulae you should remember to save some time.

Useful Summation Formulae

$$\sum_{n=1}^{N} n = \frac{N(N+1)}{2}$$

$$\sum_{n=1}^{N} n^2 = \frac{N(N+1)(2N+1)}{6}$$

$$\sum_{n=1}^{N} n^3 = \frac{N^2(N+1)^2}{4}$$

For example, the summations in examples 11.1-1, 11.1-3, and 11.1-4 can be evaluated as follows:

$$\sum_{n=1}^{10} n = \frac{10(10+1)}{2} = 55,$$

$$\sum_{n=1}^{4} n^2 = \frac{4(4+1)(2\times 4+1)}{6} = 30,$$

and

$$\sum_{n=1}^{8} n^3 = \frac{8^2(8+1)^2}{4} = 1296.$$

These formulae are especially useful when you need to evaluate a summation that has a really large upper limit such as

$$\sum_{n=1}^{100} n^2 = \frac{100(100+1)(2\times 100+1)}{6} = 338350.$$

11.2 Introduction to Product

In this section, we will learn about the product of an expression from some value to some other value.

Product
Let $f(n)$ be an expression in terms of n. Then, the product of $f(n)$ from $n = a$ to $n = b$, where a and b are integers and $a \leq b$, is

$$\prod_{n=a}^{b} f(n) = f(a) \times f(a+1) \times f(a+2) \times \cdots \times f(b).$$

n is called the index of the product. $n = a$ is the lower limit, and $n = b$ is the upper limit of the product.

Now let us do some examples.

Example 11.2-1: Find $\prod_{n=3}^{6} n$.

$$\prod_{n=3}^{6} n = 3 \times 4 \times 5 \times 6 = 360.$$

Example 11.2-2: Find $\prod_{n=2}^{7} \sqrt{n}$.

$$\prod_{n=2}^{7} \sqrt{n} = \sqrt{2} \times \sqrt{3} \times \sqrt{4} \times \sqrt{5} \times \sqrt{6} \times \sqrt{7} = 12\sqrt{35}.$$

Example 11.2-3: Find $\prod_{n=1}^{7} e^n$.

$$\prod_{n=1}^{7} e^n = e \times e^2 \times e^3 \times e^4 \times e^5 \times e^6 \times e^7 = e^{28}.$$

The product also has some properties that are similar to those of the summation.

Properties of Product
Let $f(n)$ and $g(n)$ be expressions in terms of n. Then,

$$\prod_{n=a}^{b} f(n)g(n) = \prod_{n=a}^{b} f(n) \prod_{n=a}^{b} g(n).$$

If $b \geq a \geq c$ for some integers a, b, and c, and $f(n)$ is an expression in terms of n, then

$$\prod_{n=a}^{b} f(n) = \prod_{n=c}^{b} f(n) \div \prod_{n=c}^{a} f(n).$$

If $b \geq a$ for some integers a and b, c is an integer, and $f(n)$ is an expression in terms of n, then we can re-index a product, or change the limits of the index of a product, as follows:

$$\prod_{n=a}^{b} f(n) = \prod_{n=a+c}^{b+c} f(n-c).$$

Also, by using the rules of logarithms and exponents, we have the following relations between the product and the summation.

Relations between Summation and Product
Let a and b be integers such that $b \geq a$, c be any non-zero complex number, and $f(n)$ be an expression in terms of n. Then,

$$\log_c \left(\prod_{n=a}^{b} f(n) \right) = \sum_{n=a}^{b} \log_c (f(n))$$

if

$$\prod_{n=a}^{b} f(n) \neq 0,$$

and

$$c^{\sum_{n=a}^{b} f(n)} = \prod_{n=a}^{b} c^{f(n)}.$$

11.3 Arithmetic Sequence

In this section, we will learn about a type of sequence called arithmetic sequence. A sequence is basically a list of numbers, and an arithmetic sequence is a sequence in which the difference between consecutive numbers (or terms) is constant. This constant difference is called the "common difference".

Arithmetic Sequence
An arithmetic sequence is a sequence $\{a_n\}$ such that

$$a_n = a_1 + d \times (n-1)$$

where a_1 is the first term, and d is the common difference of the arithmetic sequence.

For example,

$$2, 5, 8, 11, 14, 17, \cdots$$

is an arithmetic sequence with $a_1 = 2$ and $d = 3$.

Example 11.3-1: Given that $\{a_n\}$ is an arithmetic sequence with the first term $a_1 = 4$ and the common difference $d = 6$, find the 20th term a_{20} of the arithmetic sequence $\{a_n\}$.

$$a_{20} = a_1 + d \times (20 - 1) = 4 + 6 \times 19 = 118.$$

Example 11.3-2: The 10th term and the 30th term of an arithmetic sequence is 25 and 105, respectively. Find the first term of the arithmetic sequence.
The 10th term is 25:

$$a_{10} = a_1 + 9d = 25. \tag{11.1}$$

The 30th term is 105:

$$a_{30} = a_1 + 29d = 105. \tag{11.2}$$

Subtracting (11.1) from (11.2),

$$a_{30} - a_{10} = 20d = 80.$$

So, the common difference is $d = 4$, and we have

$$a_{10} = a_1 + 9 \times 4 = 25.$$

Solving for a_1, the first term of the arithmetic sequence is $a_1 = -11$.

Next, let us discuss arithmetic series, which is the sum of terms of an arithmetic sequence. Let us take a look at how to evaluate the sum of the first N terms of an arithmetic sequence, i.e. how to evaluate

$$\sum_{n=1}^{N} a_n$$

where $\{a_n\}$ is an arithmetic sequence. Using the definition of arithmetic sequence,

$$\sum_{n=1}^{N} a_n = a_1 + (a_1 + d) + (a_1 + 2d) + \cdots + (a_1 + (N-1)d)\,. \tag{11.3}$$

We can rewrite the sum in (11.3) in reversed order as follows:

$$\sum_{n=1}^{N} a_n = (a_1 + (N-1)d) + (a_1 + (N-2)d) + (a_1 + (N-3)d) + \cdots + a_1. \tag{11.4}$$

Adding (11.3) and (11.4),

$$2\sum_{n=1}^{N} a_n = \underbrace{(2a_1 + (N-1)d) + \cdots + (2a_1 + (N-1)d)}_{N \text{ terms}}$$
$$= N\left(2a_1 + (N-1)d\right).$$

So, the sum of the first N terms of an arithmetic sequence $\{a_n\}$ is

$$\sum_{n=1}^{N} a_n = \frac{N\left(2a_1 + (N-1)d\right)}{2}.$$

Also, since $a_1 + (N-1)d = a_N$,

$$\sum_{n=1}^{N} a_n = \frac{N\left(a_1 + a_N\right)}{2}.$$

Sum of the First N Terms of an Arithmetic Sequence
For an arithmetic sequence $\{a_n\}$,

$$\sum_{n=1}^{N} a_n = \frac{N\left(a_1 + a_N\right)}{2}.$$

Example 11.3-3: $3 + 8 + 13 + 18 + \cdots + 293 + 298 = ?$
Since $a_N = a_1 + (N-1)d$, the number of terms is

$$N = \frac{a_N - a_1}{d} + 1.$$

In this case, $a_1 = 3$, $a_N = 298$, and $d = 5$. So, there are

$$\frac{298 - 3}{5} + 1 = 60$$

terms. Applying the formula for the sum of the first N terms of an arithmetic sequence with $N = 60$, $a_1 = 3$, and $a_N = a_{60} = 298$,

$$3 + 8 + 13 + 18 + \cdots + 293 + 298 = \frac{60\,(3 + 298)}{2} = 9030.$$

11.4 Geometric Sequence

In this section, we will learn about another type of sequence called geometric sequence. In a geometric sequence, the ratio between consecutive numbers (or terms) is constant, and this constant ratio is called the "common ratio".

Geometric Sequence
A geometric sequence is a sequence $\{a_n\}$ such that

$$a_n = a_1 \times r^{n-1}$$

where a_1 is the first term, and r is the common ratio of the geometric sequence.

For example,

$$1, \frac{1}{2}, \frac{1}{4}, \frac{1}{8}, \frac{1}{16}, \frac{1}{32}, \cdots$$

is a geometric sequence with $a_1 = 1$ and $r = \frac{1}{2}$.

Example 11.4-1: Given that $\{a_n\}$ is a geometric sequence with the first term $a_1 = 2$ and the common ratio $r = 3$, find the 5th term a_5 of the geometric sequence $\{a_n\}$.

$$a_5 = a_1 \times r^{5-1} = 2 \times 3^4 = 162.$$

Example 11.4-2: The 3rd term and the 6th term of a geometric sequence is $\frac{1}{48}$ and $\frac{1}{3072}$, respectively. Find the first term of the geometric sequence.
The 3rd term is $\frac{1}{48}$:

$$a_3 = a_1 \times r^2 = \frac{1}{48}. \qquad (11.5)$$

The 6th term is $\frac{1}{3072}$:

$$a_6 = a_1 \times r^5 = \frac{1}{3072}. \qquad (11.6)$$

Divide (11.6) by (11.5),

$$\frac{a_6}{a_3} = r^3 = \frac{1}{64}.$$

So, the common ratio is $r = \frac{1}{4}$, and we have

$$a_3 = a_1 \times \left(\frac{1}{4}\right)^2 = \frac{1}{48}.$$

Solving for a_1, the first term of the geometric sequence is $a_1 = \frac{1}{3}$.

Next, let us discuss geometric series, which is the sum of terms of a geometric sequence. Let us take a look at how to evaluate the sum of the first N terms of a geometric sequence, i.e. how to evaluate

$$\sum_{n=1}^{N} a_1 r^{n-1}$$

or equivalently

$$\sum_{n=0}^{N-1} a_1 r^n$$

where a_1 is the first term and r is the common ratio of the geometric sequence. Note that

$$\sum_{n=0}^{N-1} a_1 r^n = a_1 + a_1 r + a_1 r^2 + \cdots + a_1 r^{N-1}. \tag{11.7}$$

Multiplying both sides of (11.7) by r,

$$r \sum_{n=0}^{N-1} a_1 r^n = a_1 r + a_1 r^2 + a_1 r^3 + \cdots + a_1 r^N. \tag{11.8}$$

Subtracting (11.8) from (11.7),

$$\begin{aligned}(1-r)\sum_{n=0}^{N-1} a_1 r^n &= a_1 - a_1 r^N \\ \sum_{n=0}^{N-1} a_1 r^n &= \frac{a_1\left(1-r^N\right)}{1-r}.\end{aligned}$$

Notice that r must not be one because the denominator $1-r$ has to be non-zero. Next, let us take a look at how to evaluate the sum of infinitely many terms of a geometric sequence, i.e. how to evaluate

$$\sum_{n=0}^{\infty} a_1 r^n.$$

First of all, what does it mean to have an infinite upper limit for the summation? It simply means that the upper limit is approaching infinity, i.e.

$$\sum_{n=0}^{\infty} a_1 r^n = \lim_{N\to\infty} \sum_{n=0}^{N-1} a_1 r^n.$$

Also, when dealing with an infinite sum like this, you need to understand the concept of convergence.

Convergence of an Infinite Sum

Let $f(n)$ be an expression in terms of n, a be an integer, and b be a finite real number. If

$$\sum_{n=a}^{\infty} f(n) = b,$$

then we say the sum converges to b. If

$$\sum_{n=a}^{\infty} f(n) = \pm\infty,$$

which means the sum does not have a finite value, then we say the sum diverges.

Now let us get back to the infinite sum of a geometric sequence's terms. We have

$$\sum_{n=0}^{\infty} a_1 r^n = \lim_{N\to\infty} \sum_{n=0}^{N-1} a_1 r^n = \lim_{N\to\infty} \frac{a_1\left(1-r^N\right)}{1-r}.$$

For $|r| > 1$,

$$\lim_{N\to\infty} r^N$$

diverges. For $|r| < 1$,

$$\lim_{N\to\infty} r^N = 0.$$

For example, take $r = 0.8$, and you can see that 0.8^{1000} is really close to 0. So, the sum

$$\sum_{n=0}^{\infty} a_1 r^n$$

converges for $|r| < 1$, and

$$\sum_{n=0}^{\infty} a_1 r^n = \lim_{N\to\infty} \frac{a_1\left(1-r^N\right)}{1-r} = \frac{a_1(1-0)}{1-r} = \frac{a_1}{1-r}.$$

Sum of a Geometric Sequence's Terms

$$\sum_{n=0}^{N-1} a_1 r^n = \frac{a_1\left(1-r^N\right)}{1-r}, r \neq 1.$$

$$\sum_{n=0}^{\infty} a_1 r^n = \frac{a_1}{1-r}, |r| < 1.$$

Example 11.4-3: $\frac{1}{2} + \frac{1}{4} + \frac{1}{8} + \frac{1}{16} + \cdots = ?$
Note that

$$\frac{1}{2}, \frac{1}{4}, \frac{1}{8}, \frac{1}{16}, \cdots$$

is a geometric sequence with the first term $a_1 = \frac{1}{2}$ and the common ratio $r = \frac{1}{2}$. So,

$$\frac{1}{2} + \frac{1}{4} + \frac{1}{8} + \frac{1}{16} + \cdots = \sum_{n=0}^{\infty} \frac{1}{2}\left(\frac{1}{2}\right)^n = \frac{\frac{1}{2}}{1 - \frac{1}{2}} = 1.$$

Also, note that this infinite sum converges because $|r| = \left|\frac{1}{2}\right| < 1$.

Now let us take a look at the product of the first N terms of a geometric sequence, i.e.

$$\prod_{n=1}^{N} a_1 r^{n-1}$$

or equivalently

$$\prod_{n=0}^{N-1} a_1 r^n.$$

We can see that

$$\begin{aligned}\prod_{n=0}^{N-1} a_1 r^n &= a_1 \times a_1 r \times a_1 r^2 \times \cdots \times a_1 r^{N-1} \\ &= \underbrace{a_1 \times a_1 \times a_1 \times \cdots \times a_1}_{N \text{ times}} \times r^{1+2+3+\cdots+N-1} \\ &= a_1{}^N r^{1+2+3+\cdots+N-1}.\end{aligned}$$

By using the formula for the sum of an arithmetic sequence's terms,

$$1 + 2 + 3 + \cdots + N - 1 = \frac{(N-1)(N-1+1)}{2} = \frac{N(N-1)}{2},$$

so

$$\prod_{n=0}^{N-1} a_1 r^n = a_1{}^N r^{\frac{N(N-1)}{2}}.$$

Product of the First N Terms of a Geometric Sequence

$$\prod_{n=0}^{N-1} a_1 r^n = a_1{}^N r^{\frac{N(N-1)}{2}}$$

Exercise Problems

1. Prove that $\sum_{n=1}^{N}(2n-1) = N^2$.

2. Evaluate $\prod_{n=1}^{50} e^{n^3}$.

3. Evaluate $\sum_{n=0}^{\infty}\left(\frac{1}{3}\right)^n$.

4. For $|r| < 1$, evaluate $\sum_{n=0}^{\infty} nr^n$. (Hint: differentiate a sum we have learned in this chapter with respect to r.)

5. For $|x| < 1$, evaluate $\prod_{n=0}^{\infty}\left(1 + x^{2^n}\right)$. (Hint: expand a few terms of the product and convert the product into a sum we have learned in this chapter.)

6.

Harmonic Numbers
The n-th harmonic number, denoted by H_n, is defined as

$$H_n = \sum_{k=1}^{n} \frac{1}{k}.$$

a) $H_1 =?$
b) $H_2 =?$
c) $H_5 =?$

Chapter 12

Taylor Series and Maclaurin Series

12.1 Taylor Series, Linear Approximation, and Quadratic Approximation

Taylor series of a function $f(x)$ is a series expansion of $f(x)$ using the derivatives of $f(x)$.

Taylor Series
Taylor series of a function $f(x)$ about a point $x = a$ is

$$\begin{aligned} f(x) &= \sum_{n=0}^{\infty} \frac{f^{(n)}(a)(x-a)^n}{n!} \\ &= f(a) + f'(a)(x-a) + \frac{1}{2!}f''(a)(x-a)^2 + \cdots . \end{aligned}$$

We can approximate the value of a function $f(x)$ near a point $x = a$ by using the first two or three terms of the Taylor series expansion of $f(x)$ about the point $x = a$. For example, if we want to approximate the value of $f(1.12)$, we would use the first two or three terms of the Taylor series expansion of $f(x)$ about the point $x = 1$ because 1.12 is near to 1.

If we use the first two terms of the Taylor series expansion of $f(x)$ to approximate, that is called linear approximation because the highest order derivative present is first derivative. If we use the first three terms, that is called quadratic approximation because the highest order derivative present is second derivative.

Linear Approximation and Quadratic Approximation
Linear approximation near $x = a$:

$$f(x) \approx f(a) + f'(a)(x-a).$$

Quadratic approximation near $x = a$:

$$f(x) \approx f(a) + f'(a)(x-a) + \frac{1}{2}f''(a)(x-a)^2.$$

Example 12.1-1: Use linear approximation to approximate $\sqrt{4.02}$.
Let $f(x) = \sqrt{x}$. Since 4.02 is near to 4, we will use linear approximation near $x = 4$:

$$f(4.02) \approx f(4) + f'(4)(4.02 - 4).$$

The derivative of $f(x)$ is

$$f'(x) = \frac{1}{2\sqrt{x}},$$

so $f'(4) = \frac{1}{4}$. Therefore, we obtain the approximation

$$\sqrt{4.02} \approx 2 + \frac{1}{4}(4.02 - 4) = 2.005.$$

Example 12.1-2: Use quadratic approximation to approximate $\ln 1.2$. Let $f(x) = \ln x$. Since 1.2 is near to 1, we will use quadratic approximation near $x = 1$:

$$f(1.2) \approx f(1) + f'(1)(1.2 - 1) + \frac{1}{2}f''(1)(1.2 - 1)^2.$$

The derivative and second derivative of $f(x)$ are

$$f'(x) = \frac{1}{x}$$

and

$$f''(x) = -\frac{1}{x^2},$$

so $f'(1) = 1$ and $f''(1) = -1$. Therefore, we obtain the approximation

$$\ln 1.2 \approx 0 + (1.2 - 1) - \frac{1}{2}(1.2 - 1)^2 = 0.18.$$

Example 12.1-3: Use linear approximation to approximate $\sqrt[3]{8.36}$. Let $f(x) = \sqrt[3]{x}$. Since 8.36 is near to 8, we will use linear approximation near $x = 8$:

$$f(8.36) \approx f(8) + f'(8)(8.36 - 8).$$

The derivative of $f(x)$ is

$$f'(x) = \frac{1}{3\sqrt[3]{x^2}},$$

so $f'(8) = \frac{1}{12}$. Therefore, we obtain the approximation

$$\sqrt[3]{8.36} \approx 2 + \frac{1}{12}(8.36 - 8) = 2.03.$$

12.2 Maclaurin Series

Maclaurin series of a function $f(x)$ is simply the Taylor series of $f(x)$ about the point $x = 0$.

Maclaurin Series

$$f(x) = \sum_{n=0}^{\infty} \frac{f^{(n)}(0)\,x^n}{n!} = f(0) + f'(0)x + \frac{f''(0)}{2!}x^2 + \frac{f^{(3)}(0)}{3!}x^3 + \cdots$$

Also, here are some basic, common Maclaurin series you should memorize. I would also recommend that you try to derive them using the definition of Maclaurin series and using what we learned about higher derivatives in chapter 10. The conditions after commas are the conditions for the series to converge. If there is no condition after a series, then that series converges for all real x.

Some Common Maclaurin Series

$$\sin x = \sum_{n=0}^{\infty} \frac{(-1)^n x^{2n+1}}{(2n+1)!} = x - \frac{x^3}{3!} + \frac{x^5}{5!} - \cdots$$

$$\cos x = \sum_{n=0}^{\infty} \frac{(-1)^n x^{2n}}{(2n)!} = 1 - \frac{x^2}{2!} + \frac{x^4}{4!} - \cdots$$

$$\tan^{-1} x = \sum_{n=0}^{\infty} \frac{(-1)^n x^{2n+1}}{2n+1} = x - \frac{x^3}{3} + \frac{x^5}{5} - \cdots, |x| \leq 1$$

$$\sinh x = \sum_{n=0}^{\infty} \frac{x^{2n+1}}{(2n+1)!} = x + \frac{x^3}{3!} + \frac{x^5}{5!} + \cdots$$

$$\cosh x = \sum_{n=0}^{\infty} \frac{x^{2n}}{(2n)!} = 1 + \frac{x^2}{2!} + \frac{x^4}{4!} + \cdots$$

$$\tanh^{-1} x = \sum_{n=0}^{\infty} \frac{x^{2n+1}}{2n+1} = x + \frac{x^3}{3} + \frac{x^5}{5} + \cdots, |x| < 1$$

$$e^x = \sum_{n=0}^{\infty} \frac{x^n}{n!} = 1 + x + \frac{x^2}{2} + \cdots$$

$$\ln(1-x) = -\sum_{n=1}^{\infty} \frac{x^n}{n} = -x - \frac{x^2}{2} - \frac{x^3}{3} - \cdots, x \in [-1, 1)$$

$$\ln(1+x) = \sum_{n=1}^{\infty} \frac{(-1)^{n+1} x^n}{n} = x - \frac{x^2}{2} + \frac{x^3}{3} - \cdots, x \in (-1, 1]$$

Let us derive one of these Maclaurin series here. For example, let us derive the Maclaurin series of $\ln(1-x)$. If we keep differentiating $f(x) = \ln(1-x)$, we would get

$$f^{(1)}(x) = -\frac{1}{1-x},$$
$$f^{(2)}(x) = -\frac{1}{(1-x)^2},$$
$$f^{(3)}(x) = -\frac{1 \times 2}{(1-x)^3},$$
$$f^{(4)}(x) = -\frac{1 \times 2 \times 3}{(1-x)^4},$$

and so on. So, we can see the general pattern that

$$f^{(n)}(x) = -\frac{(n-1)!}{(1-x)^n}$$

for any positive integer n. Substituting $x = 0$,

$$f^{(n)}(0) = -(n-1)!$$

for any positive integer n. Also, we know that $f(0) = \ln(1-0) = 0$. By definition, the Maclaurin series of a function $f(x)$ is

$$f(x) = \sum_{n=0}^{\infty} \frac{f^{(n)}(0)\, x^n}{n!} = f(0) + \sum_{n=1}^{\infty} \frac{f^{(n)}(0)\, x^n}{n!}.$$

Substituting $f(0) = 0$ and $f^{(n)}(0) = -(n-1)!$,

$$\ln(1-x) = 0 - \sum_{n=1}^{\infty} \frac{(n-1)!\, x^n}{n!}$$
$$= -\sum_{n=1}^{\infty} \frac{(n-1)!\, x^n}{n!}.$$

Now note that

$$\frac{(n-1)!}{n!} = \frac{(n-1) \times (n-2) \times \cdots \times 2 \times 1}{n \times (n-1) \times (n-2) \times \cdots \times 2 \times 1} = \frac{1}{n}.$$

So,

$$\ln(1-x) = -\sum_{n=1}^{\infty} \frac{x^n}{n}.$$

Example 12.2-1: Evaluate $\displaystyle\sum_{n=0}^{\infty} \frac{\ln^n \pi}{n!}$.

The Maclaurin series of e^x is

$$e^x = \sum_{n=0}^{\infty} \frac{x^n}{n!}.$$

Substituting $x = \ln \pi$, we obtain

$$\sum_{n=0}^{\infty} \frac{\ln^n \pi}{n!} = e^{\ln \pi} = \pi.$$

Example 12.2-2: For $|x| \leq 1$, what is the Maclaurin series of $\ln\left(1 + x^2\right)$? The Maclaurin series of $\ln(1 + x)$ is

$$\ln(1 + x) = \sum_{n=1}^{\infty} \frac{(-1)^{n+1} x^n}{n}$$

for $-1 < x \leq 1$. Substituting $x \to x^2$,

$$\ln\left(1 + x^2\right) = \sum_{n=1}^{\infty} \frac{(-1)^{n+1}\left(x^2\right)^n}{n} = \sum_{n=1}^{\infty} \frac{(-1)^{n+1} x^{2n}}{n}.$$

For $|x| \leq 1$, or $-1 \leq x \leq 1$, $0 \leq x^2 \leq 1$, so the series converges.

Example 12.2-3: Expand $\ln(1 - e^{-x})$ as a series. What is the domain of x where that series converges?
The Maclaurin series of $\ln(1 - x)$ is

$$\ln(1 - x) = -\sum_{n=1}^{\infty} \frac{x^n}{n}$$

for $-1 \leq x < 1$. Substituting $x \to e^{-x}$,

$$\ln\left(1 - e^{-x}\right) = -\sum_{n=1}^{\infty} \frac{\left(e^{-x}\right)^n}{n} = -\sum_{n=1}^{\infty} \frac{e^{-nx}}{n}. \tag{12.1}$$

For $x \in (0, \infty)$, $0 < e^{-x} < 1$. So, the series in (12.1) converges for $x \in (0, \infty)$.

12.3 An Interesting Sum

Before I end this chapter, I would like to introduce an interesting sum whose evaluation involves Maclaurin series:

$$\sum_{n=1}^{\infty} \frac{1}{n^2}.$$

To evaluate this sum, let us consider the function

$$\frac{\sin x}{x}.$$

The Maclaurin series of $\sin x$ is

$$\sin x = \sum_{n=0}^{\infty} \frac{(-1)^n x^{2n+1}}{(2n+1)!}$$

for all real x. Assuming $x \neq 0$, dividing both sides by x,

$$\frac{\sin x}{x} = \sum_{n=0}^{\infty} \frac{(-1)^n x^{2n}}{(2n+1)!}. \tag{12.2}$$

Now let us factor out the function $\frac{\sin x}{x}$ using its zeros. Since $\frac{\sin x}{x}$ has zeros at $x = \pm n\pi$ where n is a positive integer and

$$\lim_{x \to 0} \frac{\sin x}{x} = 1$$

(as seen in section 2.6 of chapter 2), $\frac{\sin x}{x}$ can be factored out as

$$\begin{aligned}
\frac{\sin x}{x} &= \left(1 - \frac{x}{\pi}\right)\left(1 + \frac{x}{\pi}\right)\left(1 - \frac{x}{2\pi}\right)\left(1 + \frac{x}{2\pi}\right)\left(1 - \frac{x}{3\pi}\right)\left(1 + \frac{x}{3\pi}\right)\cdots \\
&= \left(1 - \frac{x^2}{\pi^2}\right)\left(1 - \frac{x^2}{4\pi^2}\right)\left(1 - \frac{x^2}{9\pi^2}\right)\cdots \\
&= \prod_{n=1}^{\infty}\left(1 - \frac{x^2}{n^2\pi^2}\right). \qquad (12.3)
\end{aligned}$$

From (12.2) and (12.3), we get

$$\prod_{n=1}^{\infty}\left(1 - \frac{x^2}{n^2\pi^2}\right) = \sum_{n=0}^{\infty} \frac{(-1)^n x^{2n}}{(2n+1)!}. \tag{12.4}$$

The sum on the RHS of (12.4) is

$$\sum_{n=0}^{\infty} \frac{(-1)^n x^{2n}}{(2n+1)!} = 1 - \frac{1}{3!}x^2 + \frac{1}{5!}x^4 - \cdots,$$

so the coefficient of the x^2 term is $-\frac{1}{3!}$. If we expand the terms of the product on the LHS of (12.4), the coefficient of the x^2 term is

$$-\frac{1}{1^2\pi^2} - \frac{1}{2^2\pi^2} - \frac{1}{3^2\pi^2} - \frac{1}{4^2\pi^2} - \cdots.$$

By comparing the coefficients, we obtain

$$\begin{aligned}
-\frac{1}{\pi^2} - \frac{1}{2^2\pi^2} - \frac{1}{3^2\pi^2} - \frac{1}{4^2\pi^2} - \cdots &= -\frac{1}{3!} \\
\frac{1}{1^2} + \frac{1}{2^2} + \frac{1}{3^2} + \frac{1}{4^2} + \cdots &= \frac{\pi^2}{6}.
\end{aligned}$$

Therefore,

$$\sum_{n=1}^{\infty} \frac{1}{n^2} = \frac{\pi^2}{6}.$$

The sum we just evaluated is related to a special function called Riemann zeta function.

Riemann zeta function

$$\zeta(s) = \sum_{n=1}^{\infty} \frac{1}{n^s}$$

The letter ζ is a Greek letter called zeta. The sum we evaluated in this section is $\zeta(2)$.

Exercise Problems

1. Let $f(x) = x^{x^x}$. Use linear approximation to approximate $f(1.1)$.
2. Use quadratic approximation to approximate $\sqrt{4.8}$.
3. Find the value of $\frac{\pi^2}{3!} - \frac{\pi^4}{5!} + \frac{\pi^6}{7!} - \frac{\pi^8}{9!} + \cdots$.
4. Evaluate $\sum_{n=1}^{\infty} \frac{(-1)^n}{n}$.
5. Evaluate $\sum_{n=0}^{\infty} \frac{(-1)^n}{2n+1}$.

Chapter 13

Euler's Formula with Trigonometric Functions and Hyperbolic Functions

13.1 Euler's Formula

Until now, we have always stayed in the realm of real numbers. In this chapter, we will make a short journey into the realm of complex numbers. Euler's formula is an interesting formula which shows the relation between complex exponents and trigonometric functions sine and cosine.

> **Euler's Formula**
> $$e^{ix} = \cos x + i \sin x$$
> where $i = \sqrt{-1}$ and x is any complex number.

There are many proofs for this formula, but I will show you one simple proof using the Maclaurin series we just learned in the last chapter. So, now you should realize that I kind of lied to you when I said that the series without any conditions converge for all real numbers. In fact, they converge for all complex numbers - which, of course, include real numbers.

The Maclaurin series of e^x is

$$e^x = \sum_{n=0}^{\infty} \frac{x^n}{n!} = 1 + x + \frac{x^2}{2!} + \frac{x^3}{3!} + \frac{x^4}{4!} + \frac{x^5}{5!} + \frac{x^6}{6!} + \frac{x^7}{7!} + \cdots.$$

Substituting $x \to ix$,

$$e^{ix} = \sum_{n=0}^{\infty} \frac{i^n x^n}{n!} = 1 + ix + \frac{i^2x^2}{2!} + \frac{i^3x^3}{3!} + \frac{i^4x^4}{4!} + \frac{i^5x^5}{5!} + \frac{i^6x^6}{6!} + \frac{i^7x^7}{7!} + \cdots.$$

Notice that

$$\begin{aligned} i^0 &= 1, \\ i^1 &= i, \\ i^2 &= -1, \\ i^3 &= -i, \end{aligned}$$

and this cycle repeats for every four powers of i. So, we obtain

$$\begin{aligned} e^{ix} &= 1 + ix - \frac{x^2}{2!} - \frac{ix^3}{3!} + \frac{x^4}{4!} + \frac{ix^5}{5!} - \frac{x^6}{6!} - \frac{ix^7}{7!} + \cdots \\ &= \left(1 - \frac{x^2}{2!} + \frac{x^4}{4!} - \frac{x^6}{6!} + \cdots\right) + i\left(x - \frac{x^3}{3!} + \frac{x^5}{5!} - \frac{x^7}{7!} + \cdots\right). \end{aligned}$$

The Maclaurin series of $\sin x$ and $\cos x$ are

$$\sin x = \sum_{n=0}^{\infty} \frac{(-1)^n x^{2n+1}}{(2n+1)!} = x - \frac{x^3}{3!} + \frac{x^5}{5!} - \frac{x^7}{7!} + \cdots$$

and

$$\cos x = \sum_{n=0}^{\infty} \frac{(-1)^n x^{2n}}{(2n)!} = 1 - \frac{x^2}{2!} + \frac{x^4}{4!} - \frac{x^6}{6!} + \cdots .$$

Therefore, we obtain

$$e^{ix} = \cos x + i \sin x.$$

There is also another proof for this formula using ordinary differential equations, but it is beyond the scope of this book.

13.2 Non-High-School Definitions of Sine and Cosine

From the last section, we have

$$e^{ix} = \cos x + i \sin x. \tag{13.1}$$

Substituting $x \to -x$,

$$e^{-ix} = \cos(-x) + i \sin(-x) = \cos x - i \sin x \tag{13.2}$$

since $\cos x$ is an even function and $\sin x$ is an odd function.

Adding (13.1) and (13.2),

$$e^{ix} + e^{-ix} = 2\cos x,$$

so

$$\cos x = \frac{e^{ix} + e^{-ix}}{2}.$$

Subtracting (13.2) from (13.1),

$$e^{ix} - e^{-ix} = 2i \sin x,$$

so

$$\sin x = \frac{e^{ix} - e^{-ix}}{2i}.$$

Complex Definitions of Sine and Cosine

$$\sin x = \frac{e^{ix} - e^{-ix}}{2i}$$
$$\cos x = \frac{e^{ix} + e^{-ix}}{2}$$

With the above definitions of sine and cosine, we can extend the sine and cosine functions to all complex numbers. It is easy to derive the complex definitions for other trigonometric functions by using the definitions of other trigonometric functions in terms of sine and cosine, e.g.

$$\tan x = \frac{\sin x}{\cos x}.$$

Also, with the above complex definitions of $\sin x$ and $\cos x$, we can easily see that $\frac{d}{dx}(\sin x) = \cos x$ and that $\frac{d}{dx}(\cos x) = -\sin x$. For example,

$$\begin{aligned}
\frac{d}{dx}(\sin x) &= \frac{d}{dx}\left(\frac{e^{ix} - e^{-ix}}{2i}\right) \\
&= \frac{1}{2i}\frac{d}{dx}\left(e^{ix}\right) - \frac{1}{2i}\frac{d}{dx}\left(e^{-ix}\right) \\
&= \frac{1}{2i} \times ie^{ix} - \frac{1}{2i} \times (-i)e^{-ix} \\
&= \frac{e^{ix} + e^{-ix}}{2} \\
&= \cos x.
\end{aligned}$$

Example 13.2-1: $\sin(i) = ?$
By definition,

$$\sin x = \frac{e^{ix} - e^{-ix}}{2i}.$$

Substituting $x = i$,

$$\sin(i) = \frac{e^{i^2} - e^{-i^2}}{2i}.$$

Since $i^2 = -1$, our final answer is

$$\sin(i) = \frac{e^{-1} - e}{2i}.$$

13.3 Relations between Hyperbolic Functions and Trigonometric Functions

You might have wondered, "What was the whole purpose of digging into the complex realm of sine and cosine?" Well, it is to show the nice connections between trigonometric functions and hyperbolic functions, which we learned in chapter 1. Do you not see that the complex definition of sine looks similar to the definition of hyperbolic sine? Let us explore a little more.

By definition,

$$\sinh(x) = \frac{e^x - e^{-x}}{2}.$$

Substituting $x \to ix$,

$$\sinh(ix) = \frac{e^{ix} - e^{-ix}}{2}.$$

Multiplying both the numerator and the denominator by i,

$$\sinh(ix) = i\frac{e^{ix} - e^{-ix}}{2i} = i\sin(x).$$

By definition,

$$\sin(x) = \frac{e^{ix} - e^{-ix}}{2i}.$$

Substituting $x \to ix$,

$$\sin(ix) = \frac{e^{i^2x} - e^{-i^2x}}{2i}.$$

Multiplying both the numerator and the denominator by i,

$$\sin(ix) = i\frac{e^{i^2x} - e^{-i^2x}}{2i^2}.$$

Since $i^2 = -1$,

$$\sin(ix) = i\frac{e^{x} - e^{-x}}{2} = i\sinh(x).$$

By definition,

$$\cosh(x) = \frac{e^{x} + e^{-x}}{2}.$$

Substituting $x \to ix$,

$$\cosh(ix) = \frac{e^{ix} + e^{-ix}}{2} = \cos(x).$$

By definition,

$$\cos(x) = \frac{e^{ix} + e^{-ix}}{2}.$$

Substituting $x \to ix$,

$$\cos(ix) = \frac{e^{i^2x} + e^{-i^2x}}{2}.$$

Since $i^2 = -1$,

$$\cos(ix) = \frac{e^{x} + e^{-x}}{2} = \cosh(x).$$

Thus, we have derived the relations between sine and hyperbolic sine and between cosine and hyperbolic cosine. It is easy to derive the relations between other hyperbolic and trigonometric functions by using their definitions in terms of $\sinh x$, $\cosh x$, $\sin x$, and $\cos x$, e.g.

$$\tanh x = \frac{\sinh x}{\cosh x}.$$

To summarize, we have the following relations between hyperbolic functions and trigonometric functions.

Relations between Hyperbolic and Trigonometric Functions

$$\begin{aligned} \sinh(ix) &= i\sin(x) \\ \sin(ix) &= i\sinh(x) \\ \cosh(ix) &= \cos(x) \\ \cos(ix) &= \cosh(x) \end{aligned}$$

Now recall the even-odd decomposition of e^x which we discussed in section 1.2 of chapter 1:

$$e^x = \cosh(x) + \sinh(x),$$

where $\cosh(x)$ is the even part and $\sinh(x)$ is the odd part. Substituting $x \to ix$,

$$\begin{aligned} e^{ix} &= \cosh(ix) + \sinh(ix) \\ &= \cos(x) + i\sin(x), \end{aligned}$$

which is the Euler's formula we just discussed in section 13.1. So, here is an interpretation of Euler's formula in the sense of even-odd decomposition: $\cos(x)$ is the even part of e^{ix}, and $i\sin(x)$ is the odd part.

Exercise Problems

1. $e^{i\pi} = ?$

2. $e^{\frac{i\pi}{2}} = ?$

3. $\dfrac{\tan(ix)}{\tanh(x)} = ?$

Chapter 14

Introduction to Partial Derivatives

14.1 Introduction

So far in this book, we have learned the type of derivatives called ordinary derivatives, which are derivatives of single variable functions. In this chapter, I will introduce you to another type of derivatives called partial derivatives, which are derivatives of multivariable functions, just to give you a taste of what you can expect if you continue learning calculus at higher levels.

To keep things simple, we will only deal with two-variable functions, namely $f(x, y)$. However, the concept of partial derivatives can be extended to functions with more than two variables. First of all, what are two-variable functions? Well, think back to single variable functions $f(x)$ first. Single variable functions $f(x)$ are functions whose inputs are one-dimensional x and whose outputs lay on the second dimension y. Similarly, two-variable functions $f(x, y)$ are functions whose inputs are two-dimensional (x, y) and whose outputs lay on the third dimension z. For example, the picture on the cover of this book is the graph of a two-variable function $f(x, y) = x^2 + y^2$, which has a special name called paraboloid, and it is basically the 3D version of a parabola.

Now let us move on to the main purpose of this section, which is discussing the notations and meanings for partial derivatives. We will learn how to calculate partial derivatives in the next section.

You can see that a two-variable function $f(x, y)$ obviously has two variables with respect to which we can differentiate, which means we can have a partial derivative of $f(x, y)$ with respect to x and a partial derivative of $f(x, y)$ with respect to y. In the context of single variable functions, the derivative of $f(x)$ with respect to x, $f'(x)$, is the slope of the graph of $f(x)$ at a certain point, or the rate of change of $f(x)$, i.e. the amount of change in $f(x)$ caused by a change in x. Similarly, in the context of two-variable functions, the partial derivative of $f(x, y)$ with respect to x is the amount of change in $f(x, y)$ caused by a change in x, and the partial derivative of $f(x, y)$ with respect to y is the amount of change in $f(x, y)$ caused by a change in y.

There are two ways to denote partial derivatives. The partial derivative of $f(x, y)$ with respect to x can be denoted as

$$f_x(x, y)$$

or

$$\frac{\partial}{\partial x} f(x, y),$$

and the partial derivative of $f(x, y)$ with respect to y can be denoted as

$$f_y(x, y)$$

or

$$\frac{\partial}{\partial y} f(x, y).$$

What about second partial derivatives? For a two-variable function $f(x, y)$, there are four second partial derivatives: second partial derivative with respect to x twice, second partial derivative first with respect to x and then with respect to y, second partial derivative first with respect to y and then with respect to x, and second partial derivative with respect to y twice.

The second partial derivative with respect to x twice can be denoted as

$$f_{xx}(x, y)$$

or

$$\frac{\partial}{\partial x}\left(\frac{\partial}{\partial x} f(x, y)\right) = \frac{\partial^2}{\partial x \partial x} f(x, y).$$

The second partial derivative first with respect to x and then with respect to y can be denoted as

$$f_{xy}(x, y)$$

or

$$\frac{\partial}{\partial y}\left(\frac{\partial}{\partial x} f(x, y)\right) = \frac{\partial^2}{\partial y \partial x} f(x, y).$$

The second partial derivative first with respect to y and then with respect to x can be denoted as

$$f_{yx}(x, y)$$

or

$$\frac{\partial}{\partial x}\left(\frac{\partial}{\partial y} f(x, y)\right) = \frac{\partial^2}{\partial x \partial y} f(x, y).$$

The second partial derivative with respect to y twice can be denoted as

$$f_{yy}(x, y)$$

or

$$\frac{\partial}{\partial y}\left(\frac{\partial}{\partial y} f(x, y)\right) = \frac{\partial^2}{\partial y \partial y} f(x, y).$$

Also, if the second partial derivatives $f_{xy}(x, y)$ and $f_{yx}(x, y)$ are continuous, then we have the following equality:

$$f_{xy}(x, y) = f_{yx}(x, y).$$

In general, if we use the subscript notation for partial derivatives, the order of the variables with respect to which we differentiate is from left to right. For example,

$$f_{xyyxyx}(x, y)$$

is the sixth partial derivative of $f(x, y)$ first with respect to x, second with respect to y, third with respect to y, fourth with respect to x, fifth with respect to y, and then sixth with respect to x. If we use the fraction notation for partial derivatives, the order of the variables with respect to which we differentiate is from right to left. For example,

$$\frac{\partial^4}{\partial x \partial y \partial x \partial x} f(x, y)$$

is the fourth derivative of $f(x, y)$ first with respect to x, second with respect to x, third with respect to y, and then fourth with respect to x.

14.2 Formal Definitions and How to Calculate

Let us first take a look at the formal definitions of partial derivatives.

Formal Definitions of Partial Derivatives

$$\frac{\partial}{\partial x} f(x, y) = \lim_{h \to 0} \frac{f(x + h, y) - f(x, y)}{h}$$
$$\frac{\partial}{\partial y} f(x, y) = \lim_{h \to 0} \frac{f(x, y + h) - f(x, y)}{h}$$

We can see that the formal definitions for partial derivatives are very similar to the formal definition for ordinary derivatives. Also, from the definitions, we can see that nothing happens to y when we differentiate $f(x, y)$ with respect to x, and nothing happens to x when we differentiate $f(x, y)$ with respect to y.

So, when we calculate the partial derivatives with respect to a variable, we treat the other variable as a constant and calculate the partial derivatives just like how we would calculate the ordinary derivatives. For example, if we calculate the partial derivative of $f(x, y)$ with respect to x, we treat y as a constant and differentiate with respect to x just like how we would find ordinary derivatives of a single variable function $f(x)$. Now let us do some examples to familiarize ourselves with partial derivatives.

Example 14.2-1: For $f(x,y) = xy$, find $f_x(x,y)$ and $f_y(x,y)$.
When we differentiate with respect to x, we treat y as a constant. So,

$$f_x(x,y) = \frac{\partial}{\partial x}(xy) = y\frac{\partial}{\partial x}(x).$$

By using the power rule of ordinary derivatives, we have

$$\frac{\partial}{\partial x}(x) = 1,$$

so

$$f_x(x,y) = y.$$

Similarly, we also have $f_y(x,y) = x$.

Example 14.2-2: Find $\frac{\partial}{\partial y}\left(y^{\sin x}\right)$.
When we differentiate with respect to y, we treat x as a constant, so $\sin x$ is a constant. Then, using the power rule, we have

$$\frac{\partial}{\partial y}\left(y^{\sin x}\right) = \sin(x) \times y^{\sin(x)-1}.$$

Example 14.2-3: Find $\frac{\partial^2}{\partial y \partial x}(\sin x \cos y)$.
The notation

$$\frac{\partial^2}{\partial y \partial x}$$

means we are taking second partial derivative first with respect to x and then with respect to y. When we differentiate with respect to x, we treat y as a constant, so $\cos y$ is a constant:

$$\frac{\partial}{\partial x}(\sin x \cos y) = \cos(y)\frac{\partial}{\partial x}(\sin x).$$

Since the derivative of $\sin x$ is $\cos x$,

$$\frac{\partial}{\partial x}(\sin x \cos y) = \cos x \cos y.$$

Now we need to differentiate the expression above with respect to y:

$$\frac{\partial^2}{\partial y \partial x}(\sin x \cos y) = \frac{\partial}{\partial y}(\cos x \cos y).$$

When we differentiate with respect to y, we treat x as a constant, so $\cos x$ is a constant:

$$\frac{\partial}{\partial y}(\cos x \cos y) = \cos(x)\frac{\partial}{\partial y}(\cos y).$$

Since the derivative of $\cos y$ is $-\sin y$,

$$\frac{\partial}{\partial y}\left(\cos x \cos y\right) = -\cos x \sin y.$$

Therefore, our final answer is

$$\frac{\partial^2}{\partial y \partial x}\left(\sin x \cos y\right) = -\cos x \sin y.$$

Example 14.2-4: Find $\frac{\partial^2}{\partial x \partial y}\left(x^y\right)$.

The notation

$$\frac{\partial^2}{\partial x \partial y}$$

means we are taking second partial derivative first with respect to y and then with respect to x. When we differentiate with respect to y, we treat x as a constant. Then, using the formula for derivatives of exponential functions, we obtain

$$\frac{\partial}{\partial y}\left(x^y\right) = x^y \ln x.$$

Now we need to differentiate the expression above with respect to x:

$$\frac{\partial^2}{\partial x \partial y}\left(x^y\right) = \frac{\partial}{\partial x}\left(x^y \ln x\right).$$

When we differentiate with respect to x, we treat y as a constant. Applying the product rule,

$$\frac{\partial}{\partial x}\left(x^y \ln x\right) = \frac{\partial}{\partial x}\left(x^y\right) \ln x + x^y \frac{\partial}{\partial x}(\ln x).$$

Since

$$\frac{\partial}{\partial x}\left(x^y\right) = yx^{y-1}$$

and

$$\frac{\partial}{\partial x}(\ln x) = \frac{1}{x},$$

we have

$$\frac{\partial}{\partial x}\left(x^y \ln x\right) = yx^{y-1} \ln x + x^{y-1}.$$

Therefore, our final answer is

$$\frac{\partial^2}{\partial x \partial y}\left(x^y\right) = yx^{y-1} \ln x + x^{y-1}.$$

Example 14.2-5: Find $\dfrac{\partial}{\partial y}\left(\dfrac{y^x}{\sin y}\right)$.
When we differentiate with respect to y, we treat x as a constant. Applying the quotient rule,

$$\frac{\partial}{\partial y}\left(\frac{y^x}{\sin y}\right) = \frac{\frac{\partial}{\partial y}(y^x)\sin y - y^x \frac{\partial}{\partial y}(\sin y)}{\sin^2 y}.$$

Since

$$\frac{\partial}{\partial y}(y^x) = xy^{x-1}$$

and

$$\frac{\partial}{\partial y}(\sin y) = \cos y,$$

we obtain

$$\frac{\partial}{\partial y}\left(\frac{y^x}{\sin y}\right) = \frac{xy^{x-1}\sin y - y^x \cos y}{\sin^2 y}.$$

14.3 Multivariable Chain Rule

As we saw in examples 14.2-4 and 14.2-5, the product rule and the quotient rule for partial derivatives are the same as those for ordinary derivatives, which we learned in chapter 4. However, the chain rule for partial derivatives is a little different from the chain rule for ordinary derivatives. In this section, we will learn about the chain rule for partial derivatives.

Multivariable Chain Rule
For any function $f(x, y)$, let $x = x(u, t)$ and $y = y(u, t)$. Then,

$$\frac{\partial f}{\partial u} = \frac{\partial f}{\partial x} \times \frac{\partial x}{\partial u} + \frac{\partial f}{\partial y} \times \frac{\partial y}{\partial u},$$

and similarly

$$\frac{\partial f}{\partial t} = \frac{\partial f}{\partial x} \times \frac{\partial x}{\partial t} + \frac{\partial f}{\partial y} \times \frac{\partial y}{\partial t}.$$

Example 14.3-1: Let $f(x, y) = x^2 + y^2$, $x = ut$, and $y = u + t$. Find $\dfrac{\partial f}{\partial u}$.
Applying the chain rule,

$$\frac{\partial f}{\partial u} = \frac{\partial f}{\partial x} \times \frac{\partial x}{\partial u} + \frac{\partial f}{\partial y} \times \frac{\partial y}{\partial u}. \tag{14.1}$$

When we differentiate with respect to x, we treat y as a constant, so y^2 is a constant:

$$\frac{\partial f}{\partial x} = \frac{\partial}{\partial x}\left(x^2 + y^2\right) = \frac{\partial}{\partial x}\left(x^2\right) + \frac{\partial}{\partial x}\left(y^2\right) = 2x + 0 = 2x. \tag{14.2}$$

Similarly,

$$\frac{\partial f}{\partial y} = 2y. \tag{14.3}$$

When we differentiate with respect to u, we treat t as a constant:

$$\frac{\partial x}{\partial u} = \frac{\partial}{\partial u}(ut) = t\frac{\partial}{\partial u}(u) = t, \tag{14.4}$$

and

$$\frac{\partial y}{\partial u} = \frac{\partial}{\partial u}(u + t) = \frac{\partial}{\partial u}(u) + \frac{\partial}{\partial u}(t) = 1 + 0 = 1. \tag{14.5}$$

Substituting (14.2), (14.3), (14.4), and (14.5) into (14.1),

$$\frac{\partial f}{\partial u} = 2xt + 2y.$$

Also, since $x = ut$ and $y = u + t$,

$$\frac{\partial f}{\partial u} = 2ut^2 + 2u + 2t.$$

Before I end this chapter, there is still some final note I need to make about partial derivatives. There is much more about partial derivatives and their applications. However, I will not cover those topics in this book. This chapter is only intended to be a brief introduction to partial derivatives to give you a taste of what you will learn if you continue to higher levels of calculus. Of course, you can learn more about partial derivatives on your own if you want to.

Exercise Problems

1. Evaluate $\frac{\partial}{\partial x}(\sinh y \tanh x)$.
2. Evaluate $\frac{\partial^2}{\partial x \partial y}(\sin x \cos y)$.
3. Evaluate $\frac{\partial^2}{\partial y \partial x}(x^y)$.
4. Evaluate $\frac{\partial^2}{\partial y \partial y}(\cosh^y(x))$.
5. Let $f(x, y) = x^2 + y^2$, $x = ut$, and $y = u + t$. Find $\frac{\partial f}{\partial t}$.

Answers to Exercise Problems

Chapter 1

1. a) $\sqrt{5}$ b) $4\sqrt{5}$ c) 9

2. $-\dfrac{\sqrt{2}}{2}$

3. $\dfrac{1}{2}\ln 2$

4. $\ln\left(2+\sqrt{3}\right)$

5. $\ln\left(4+\sqrt{17}\right)$

Chapter 2

1. 4

2. a) 0 b) $(-1)^n$

3. $g(x) = x^3$ and $h(x) = -x^3$

4. 0

5. 1

Chapter 3

1. Solution to this will be available in the solution manual

2. Solution to this will be available in the solution manual

3. $\dfrac{1}{3\sqrt[3]{x^2}}$

4. $5^{x \ln 5} \ln^2 5$

5. $\dfrac{1}{\ln 6}$

6. 0

Chapter 4

1. 2

2. $-\dfrac{\ln 6}{x \ln^2 x}$

3. $f'(x) = \dfrac{1}{x}$; $f'(-1) = -1$

4. $f'(x) = \dfrac{3x^5}{|x^3|}$; $f'(-2) = -12$

5. $\lambda = 8$

6. $b = \dfrac{1+\sqrt{5}}{2}$ or $b = \dfrac{1-\sqrt{5}}{2}$

Chapter 5

1. $\dfrac{xy \ln y - y^2}{xy \ln x - x^2}$

2. $9x^4 g\left(x^6\right) + 6xh\left(x^3\right)$

3. $\dfrac{-\tau}{\eta^3}$

4. 0

5. Solution to this will be available in the solution manual

6. Solution to this will be available in the solution manual

7. $\dfrac{W(x)}{xW(x) + x}$

Chapter 6

1.
   a) Increasing
   b) Decreasing
   c) $x = e$; a maximum

2.
   a) $x = -\sqrt{\frac{3}{10}}$, $x = 0$, and $x = \sqrt{\frac{3}{10}}$
   b) Concave down on $\left(-\infty, -\sqrt{\frac{3}{10}}\right)$, concave up on $\left(-\sqrt{\frac{3}{10}}, 0\right)$, concave down on $\left(0, \sqrt{\frac{3}{10}}\right)$, and concave up on $\left(\sqrt{\frac{3}{10}}, \infty\right)$

Chapter 7

1. 0

2. $\frac{1}{2}$

3. $-\frac{1}{2}$

4. 0

5. a) $\frac{2}{\sqrt{\pi}}$ b) 1

Chapter 8

1. 24

2. 0

3. $\ln\left(\sqrt{2\pi}\right)$

Chapter 9

1. 0

2. 2

3. 0.567

4. $\{x\}$

5.
   a)Solution to this will be available in the solution manual
   b)Solution to this will be available in the solution manual

6. a) 30 b) 20

Chapter 10

1. $32e^2$

2. 3

3. $\frac{1}{2}\left[8^n \sin\left(8x+\frac{n\pi}{2}\right)+2^n \sin\left(2x+\frac{n\pi}{2}\right)\right]$

4. $ne^x + xe^x$

5. 124

Chapter 11

1. Solution to this will be available in the solution manual

2. $e^{1625625}$

3. $\frac{3}{2}$

4. $\frac{r}{(1-r)^2}$

5. $\frac{1}{1-x}$

6. a) 1 b) $\frac{3}{2}$ c) $\frac{137}{60}$

Chapter 12

1. $f(1.1) \approx 1.1$

2. $\sqrt{4.8} \approx 2.19$

3. 1

4. $-\ln 2$

5. $\frac{\pi}{4}$

Chapter 13

1. -1

2. i

3. i

Chapter 14

1. $\sinh y \operatorname{sech}^2 x$

2. $-\cos x \sin y$

3. $yx^{y-1} \ln x + x^{y-1}$

4. $\cosh^y(x) \ln^2(\cosh x)$

5. $2u^2t + 2u + 2t$

Sources Cited in the Book

1.1

Hyperbolic Trigonometric Functions. *Brilliant.org.* Retrieved December 21, 2019, from https://brilliant.org/wiki/hyperbolic-trigonometric-functions/

1.2

Hyperbolic Trigonometric Functions. *Brilliant.org.* Retrieved December 21, 2019, from https://brilliant.org/wiki/hyperbolic-trigonometric-functions/

Branson, J. Hyperbolic Function Identities. *hepweb.ucsd.edu.* Retrieved December 25, 2019, from https://hepweb.ucsd.edu/ph110b/110b_notes/node49.html

1.3

Weisstein, Eric W. "Inverse Hyperbolic Functions." From *MathWorld*–A Wolfram Web Resource. http://mathworld.wolfram.com/InverseHyperbolicFunctions.html

2.1

Epsilon-Delta Definition of a Limit. *Brilliant.org.* Retrieved January 7, 2020, from https://brilliant.org/wiki/epsilon-delta-definition-of-a-limit/

2.6

Squeeze Theorem. *Brilliant.org.* Retrieved January 16, 2020, from https://brilliant.org/wiki/squeeze-theorem/

Alsamraee, H. "The Limit." §1.1 in *Advanced Calculus Explored: With Applications in Physics, Chemistry, and Beyond.* Curious Math Publications, pp. 34-35, 2019.

3.10

Calculus With Inverse Trigonometric Functions. *Brilliant.org.* Retrieved February 07, 2020, from https://brilliant.org/wiki/applying-differentiation-rules-to-inverse/

3.11

Derivative of Absolute Value Function. *Proof Wiki.* Retrieved February 20, 2020, from https://proofwiki.org/wiki/Derivative_of_Absolute_Value_Function

3.12

Weierstrass Function. *Wikipedia.* Retrieved August 19, 2020, from https://en.wikipedia.org/wiki/Weierstrass_function

12.3

Euler and the Function sin(x)/x. Retrieved August 20, 2020, from https://mae.ufl.edu/~uhk/EULER-SINE.pdf

Acknowledgements

In this section, I would like to mention some people whom I would like to thank for their contributions to the process of making this book.

I owe special thanks to Professor Philip Uri Treisman, a professor of mathematics at University of Texas at Austin, who gave me some valuable advice regarding how to write a good math book.

Also, I'm grateful for my English teacher at Brentwood Christian School, Dr. Paul Robison, who has helped me fix the grammar and writing errors in this book, from the "About the Author" section to chapter 14. Whenever I sent him the pdf of the chapters in this book, he replied to my emails with helpful grammar errors suggestions very fast, usually within one to three days. Another English teacher at Brentwood Christian School, Mr. Taylor Mayfield, also fixed some grammar and writing errors in chapter 1.

Other than the grammar and writing, I appreciate Mrs. Michele Broadway, a pre-calculus teacher at Brentwood Christian School, and Mrs. Kaleen Graessle, an AP calculus teacher at Brentwood Christian School, for their feedback on the book in general.

On Instagram, I have established an edit team consisting of many members with various math backgrounds. I would like to thank those members for their helpful feedback. The following is the list of the members of the edit team in no particular order:

- Raunak Singh,
- Norman James Smith,
- Jacob Wenczkowski,
- Keyvon Rashidi,
- Jack Moffatt,
- Abdul Hannan,
- Vasilij Shikunov,
- Anton Steinfadt,

- Sai Prasad Myanamwar,
- Solden Stoll,
- Juan Diego Gamboa Aguirre,
- David Antrobius,
- Felipe Ospina Suarez,
- Dingbang Chen,
- Pablo Fernandez Golbano.

For the exercise problems in this book, other than the problems I made myself, there are also some problems suggested by Juan Diego Gamboa Aguirre.

I would also like to give thanks to the designer of the cover of this book. The cover of this book was designed by my friend who is a creative self-made designer, Minh Nguyen Chieu Phan.

Lastly, I owe my parents for their support to let me study in America. If I had not come to America, I would not have discovered my interest in math.

Reviews

"This book is suitable for curious high school students, some college students, and maybe even some curious adults. This book has a difference in a friendly, readable, and sometimes cute writing. This is truly a book written by a single author, consistent in style and contents.

There are several nice touches which I like, such as the derivation of the derivative of the exponential function on page 37. The piece about Weierstrass function [in chapter 3] and the derivation of the value of zeta function at 2 by Euler [in chapter 12] are nice and not often found in Calculus books.

On the whole, I like the book a lot, and would recommend it to many high school students. I think high school students might enjoy and use it more than other books, knowing that it is written by their peer."

Dr. Vu Quang Huynh
Dean of Faculty of Mathematics and Computer Science, Vietnam National University Ho Chi Minh City - University of Science; Head of Department of Analysis, Faculty of Mathematics and Computer Science, Vietnam National University Ho Chi Minh City - University of Science

---

"I am an associate professor of University of Science, VNU-Ho Chi Minh City, but I also teach some classes in high school. I have felt excited when knowing this book is written by a high school student. I think the author had an impressive job if compared with a lot of students of the same age in Vietnam (although he is in the US now). Not many Vietnamese students can do this job.

This book has fourteen chapters presenting basic definitions and results on calculus in one variable. The layout is very good. Many results and examples are explained very clearly. I do not know the education system in the US but as a teacher in Vietnam, I can say that the book contains difficult topics which are not taught in Vietnamese high schools, and the author presents these in a good way. However, I think this manuscript is much better if the author adds several examples and exercises."

Dr. Bien Hoang Mai
Head of Department of Algebra, Faculty of Mathematics and Computer Science, Vietnam National University Ho Chi Minh City - University of Science;
Associate Professor of Mathematics at Vietnam National University Ho Chi Minh City - University of Science

---

"*An Introduction to Calculus* provides a plethora of interesting and fun examples to work through. It is a book that illustrates many elementary concepts wonderfully and delves into them using an example-based approach. It covers a wide variety of techniques and examples, more so than a typical elementary calculus course would. This makes it a detailed yet simple book to read, perfect for a beginner aiming to master elementary calculus."

Hamza Alsamraee
Author of *Advanced Calculus Explored* and *Paradoxes*;
Admin of Instagram math page *@daily_math_*

"The book *An Introduction to Calculus: With Hyperbolic Functions, Limits, Derivatives, and More* by author Duc Van Khanh Tran refers to the theories of limits, the derivative and differential of a function of a single variable, and the partial derivative of a function of several variables in a practical and easily accessible way. Moreover, the book has covered many interesting additions in chapters 1, 8, 9. There are many relatively rich illustrative examples. The book is suitable for learners who want to research an overview of Calculus. Despite [some] shortcomings that need upgrading and editing, I am still really impressed with the author's motivation and seriousness in compiling the first book."

Dr. Triet Anh Nguyen
Head of Department of Mathematics, Mechanics, and Informatics at University of Architecture Ho Chi Minh City

---

"I am deeply impressed by [the author's] work and talent. The talent of this work lies not only in the mathematics it contains, but also in the young mathematician behind its structure and content. His exposition and synthesis of differential calculus, from its underpinnings to its applications, is delivered in a style that reveals his disposition: a personable young man with a contagious enthusiasm for sharing and learning."

Dr. Tri Cong Nguyen
PhD of Mathematics at Kanazawa University;
Vice Principal and mathematics teacher at Phan Ngoc Hien Gifted High School

---

"*An Introduction to Calculus* provides a comprehensive overview of the strategies and techniques in introductory calculus. Duc Van Khanh Tran's pedagogical language and engaging tone make the abstract concepts easy to follow. Furthermore, he includes many results nonstandard to a traditional introductory text that spark excitement at the power of math. To any student interested in exploring the ideas of calculus, this book will be hard to put down!"

Jack Moffatt
Admin of Instagram math page *@integral.fun*

"The book is well organized with concise definitions, a lot of examples with explanations, and exercise problems for further practice. I like how each worked example is explained in great detail. The topics covered are much more advanced than normal calculus textbooks. This is definitely a gift for all Math lovers to start their journey in Calculus."

Vinci Mak
Admin of Instagram math page *@vibingmath*

"I love how [the author] often used the same functions as [he] worked through the concepts and embedded and required great algebra practice along the way. [He] showed steps very clearly and labeled them clearly. Each chapter is a very manageable amount of material. And I like the progression of [this] book."

Michele Broadway
Pre-calculus teacher at Brentwood Christian School

www.ingramcontent.com/pod-product-compliance
Ingram Content Group UK Ltd.
Pitfield, Milton Keynes, MK11 3LW, UK
UKHW041638190726
13854UKWH00006B/2568

9 798736 509287